Design For Inclusion

Volume 2

Series Editor

Giuseppe Di Bucchianico, Department of Engineering and Geology,
University of Chieti-Pescara, Pescara, Italy

The book series "Design for Inclusion" publishes studies aimed at untangling the complex relationships between design and society, especially with regard to human diversity and inclusion. The series cover the classical fields of industrial design, such as product design, interior design, service design and communication design, with a special focus on their interactions with other disciplines, such as the physical and social sciences. In particular, the series reports on the latest developments, both theoretical and practical, concerning different approaches to "Design for Inclusion": Universal Design, Inclusive Design, and Design for All. It shares insights into their multiple, interdisciplinary relationships with both the human sciences and the transversal frameworks and approaches typical of Human-Centered Design (HCD) and Design for Sustainability. "Design for Inclusion" addresses a very broad readership, including designers, engineers, architects, social scientists and stakeholders, who deal with any of the topics mentioned above and their various implications. It publishes volumes fostering scientific advances, as well as books devoted to supporting education and professional training. These include monographs, edited books and conference proceeding. Outstanding PhD theses on emerging topics, if properly reworked, may also be considered for publication. This book series is published with the support of EIDD - Design for All Europe.

More information about this series at http://www.springer.com/series/16456

Rens Brankaert • Caylee Raber •
Maarten Houben • Paulina Malcolm •
Jon Hannan
Editors

Dementia Lab 2021: Supporting Ability Through Design

Proceedings of the 5th Dementia Lab Conference, D-Lab 2021, January 18–28, 2021

EIDD
DESIGN
FOR ALL
EUROPE

Springer

Editors
Rens Brankaert
Department of Industrial Design
Eindhoven University of Technology
Eindhoven, The Netherlands

School of Allied Health Professions
Fontys University of Applied Sciences
Eindhoven, The Netherlands

Maarten Houben
Department of Industrial Design
Eindhoven University of Technology
Eindhoven, The Netherlands

Jon Hannan
Health Design Lab
Emily Carr University of Art + Design
Vancouver, BC, Canada

Caylee Raber
Health Design Lab
Emily Carr University of Art + Design
Vancouver, BC, Canada

Paulina Malcolm
Health Design Lab
Emily Carr University of Art + Design
Vancouver, BC, Canada

ISSN 2662-5644 ISSN 2662-5652 (electronic)
Design For Inclusion
ISBN 978-3-030-70292-2 ISBN 978-3-030-70293-9 (eBook)
https://doi.org/10.1007/978-3-030-70293-9

This Springer imprint is published by the registered company Springer Nature Switzerland AG
The registered company address is: Gewerbestrasse 11, 6330 Cham, Switzerland

Organization

Scientific Committee

Hester van Zuthem, Active Cues Tovertafel
Rens Brankaert, Eindhoven University of Technology
Maarten Houben, Eindhoven University of Technology
Sandra Suijkerbuijk, Eindhoven University of Technology, Vilans
Caylee Raber, Emily Carr University of Art and Design
Jon Hannan, Emily Carr University of Art and Design
Lisa Boulton, Emily Carr University of Art and Design, British Columbia Institute of Technology,
Eveline Wouters, Fontys University of Applied Sciences, Tilburg University
Paul Rodgers, Lancaster University
Niels Hendriks, LUCA School of Arts
Andrea Wilkinson, LUCA School of Arts
James Hodge, Newcastle University
Dorien Oostra, Radboud University Medical Center
Claire Craig, Sheffield Hallam University
Alison Phinney, The University of British Columbia
Gloria Puurveen, The University of British Columbia
Mariko Sakamoto, The University of British Columbia
Liselore Snaphaan, Tilburg University
Sarah Foley, University College Cork, Ireland
Rita Maldonado Branco, University of Porto
David Unbehaun, University of Siegen
Sarah Campbell, University of Surrey
Gail Kenning, University of Technology Sydney
Henk-Herman Nap, Vilans

Organizing Committee

Caylee Raber, Emily Carr University of Art and Design
Jon Hannan, Emily Carr University of Art and Design
Paulina Malcolm, Emily Carr University of Art and Design, The University of British Columbia

Lisa Boulton, Emily Carr University of Art and Design, British Columbia Institute of Technology
Nadia Beyzaei, Emily Carr University of Art and Design
Garima Sood, Emily Carr University of Art and Design
Tyler Hawkins, Emily Carr University of Art and Design
Cassandra Grdovic, Emily Carr University of Art and Design
Myrna Norman, Community Partner
Mario Gregorio, Community Partner
Lynn Jackson, Community Partner
Marcia Gresch, Community Partner
Rens Brankaert, Eindhoven University of Technology
Maarten Houben, Eindhoven University of Technology
Niels Hendriks, LUCA School of Arts
Andrea Wilkinson, LUCA School of Arts

Preface

After a year of change and challenges, we are pleased to invite you to read the proceedings of Dementia Lab 2021, centred on the theme of 'Supporting Ability'. These proceedings contain papers and ideas presented virtually at the fifth installment of the conference, hosted by the Health Design Lab at Emily Carr University of Art and Design.

The Dementia Lab is an inclusive multidisciplinary community including designers, artists, researchers, health professionals, carers, and people living with dementia. This year's event and the papers contained here, highlight issues and innovations surrounding ability and disability, focusing on ways we can engage with people living with dementia that are not determined or limited by what they cannot do, but instead embrace ways in which they are able to contribute.

The papers in these proceedings speak to the themes of: (1) engagement: ethics of inclusion and shifting the culture of care to one focused on both personal independence and reconnecting with the community; (2) empowerment: new ways of designing with/for those living with dementia, embracing challenges and tackling stigma to break negative stereotypes and preconceived opinions; (3) identity: approaches to countering social isolation, retaining personhood/dignity and fostering empathy.

The ideas and conversations presented here reflect on how design and research efforts can lead the way in uncovering, supporting and enhancing the abilities of people with dementia and the community within which they live by involving them in the conversation. Little did we know when we conceived of this conference theme, that a pandemic would further challenge all of us to reconsider approaches to inclusion and participation. 'Supporting ability' in the past year has taken on new forms as designers, researchers, carers and people living with dementia have adapted to working and collaborating in new ways, facing unforeseen challenges and making the best of new opportunities. A few papers in these proceedings speak to emerging strategies for virtual engagement, furthered by live discussions and workshops in the virtual Dementia Lab 2021 event.

This year's event was organized by the Health Design Lab at Emily Carr University of Art and Design, in collaboration with the event's founders from the LUCA School of Arts (Belgium), and design researchers from Eindhoven University of Technology (The Netherlands). A core focus of the work in the Health Design Lab is centred on the use of participatory design methods to support culture change within long-term care homes, amplifying the voice and choice of people with dementia and those living in care. In adapting to the pandemic by shifting to a virtual event, the theme of inclusion for this conference is enhanced, by enabling a wider range of participants from around the world to tune in.

The international contributions towards these proceedings and this conference, enables us to see how dementia is experienced and understood from a range of perspectives. This year's submissions brought to light conversations and challenges in terms of finding the language to describe people living with dementia and the settings within which they live, as well as broadening conversations about how one identifies oneself. In fact, throughout the planning of the conference and in the conversations facilitated through the conference papers and talks, one thing that continues to emerge is the topic of language. Language encompasses value and it can be a powerful link and framing of one's identity. When it comes to dementia, there are ongoing conversations about this term and stigmas associated with it, as well as alternative terms such as Alzheimer's disease and neuro-cognitive disorder. Further, in Canada, we are working to shift away from terms such as 'resident' and 'residential care', as we are aware of the traumatic history of residential schools for Indigenous Peoples in Canada. In conversations about language, we also recognize that people generally hold multiple identities, and someone with dementia, can also identify in many other ways—including, as a researcher, a carer, a community member and a designer.

Our process for conceptualizing the conference included co-designing with a group of local community members living with dementia. Through conversations with our conference collaborators, we learned about their preferences regarding personal identification and the impacts of certain language and terminology regarding "dementia"; we learned about what makes a meaningful physical surrounding—especially while living through the pandemic; we learned how to promote cultural sensitivity within the conference, and how to foster engagement by all conference attendees. Through inviting co-designers to teach us about their skills and share their stories, we learned about their unique lived experiences, and they provided guidance on engagement guidelines for the conference—stressing the importance of learning through doing and the reciprocity of giving and gaining energy to and from others. It is in this spirit that we have endeavoured to structure the conference in a way that is inclusive and responsive to the needs of our non-academic audience and partners, and as such have created a conference zine to share additional stories to complement these proceedings.

The proceeding editors wish to thank the dedicated Scientific Committee members, the Organizing Committee, Springer and our sponsors: The Social Sciences and Humanities Research Council of Canada, Vancouver Coastal Health, Alzheimer BC and the UBC Centre for Research on Personhood in Dementia. However, first and foremost on behalf of all contributing authors we want to thank our participants and collaborators, especially those with lived experience of dementia for their contributions to the work presented here.

Vancouver, Canada

Caylee Raber
Jon Hannan

Contents

Scaling Participatory Methods for Cultural Change in Long Term Care Homes

Lisa Boulton, Caylee Raber, Nadia Beyzaei, Jo-Ann Tait, Heather Mak, Jas Gill, Monique Davidson, Vannysha Chang, Jean Chisholm, Ajra Doobenen, Laura Escueta, Morgan Martino, Nandita Ratan, Eliza Rose, and Garima Sood

Abstract

In North America, models of care within care home settings are derived from institutional hospital settings resulting in decisions over daily practices being driven by efficiency and financial accountabilities. The Culture of Care project seeks to shift the culture of care in long term care homes from an institutional model, to a social, person directed model by uplifting the perspectives of the people residing in care homes; creating the space for them to shape the future of their care. Working in collaboration with health care leaders, staff, people living in long-term care and their families, design researchers engaged participatory design research methodologies in order to foster the relationships and conversations that activate latent knowledge embodied in the people living in long-term care. Negotiating differences of approach to transformation, risk aversion in a health context and developing trust around the design research process were central to facilitating change from within the health context. Over the course of two years, the Culture of Care project created a series of workshops, media, events and toolkits that were intended to scale and connect with the people living in fifty-five care homes across the greater Vancouver area and Sunshine Coast. In this case study, we discuss how participatory design activities can be scaled to create the possibility for large organizational systems change, through deep collaboration between organizational leaders, users and external designers.

Keywords

Participatory design · Long term care · Scale

1 Shifting from Institutionalized Patient-Centered Care, to Social, Person Directed Care

Poet, Theo E.J. Wilson, describes language as "…the first form of virtual reality… a symbolic representation of the physical world" and that "through this device, we change the physical world." A patient-centred view of a person living in a care home positions them as a patient first—

L. Boulton (✉) · C. Raber · N. Beyzaei · V. Chang · J. Chisholm · A. Doobenen · L. Escueta · M. Martino · N. Ratan · E. Rose · G. Sood
Health Design Lab, Emily Carr University of Art + Design, Vancouver, BC, Canada
e-mail: raber@ecuad.ca

J.-A. Tait · H. Mak · J. Gill · M. Davidson
Vancouver Coastal Health Authority, Vancouver, BC, Canada

R. Brankaert et al. (eds.), *Dementia Lab 2021: Supporting Ability Through Design*, Design For Inclusion 2, https://doi.org/10.1007/978-3-030-70293-9_1

not as a person. Definitions used to describe how a person can be centered in their care when considering people living with Dementia (PLWD) have come to include such a wide array of approaches or values that it's meaning has become clouded [1]. Seldom does the language associated with centering PLWD speak to how the person themselves can be a contribution. Person-directed care is a designed language change intended to see people, not patients living in care homes. Person-directed care looks beyond care staff; it is looking at the people living in long-term care (LTC) to activate possibilities for agency.

In a similar manner, language used in this project to refer to LTC avoid words like "facility" and other highly medicalized or institutionalized language that positions a place where people live as anything other than "Home." Emphasizing the *social* is a call to encourage exchange, to connect with storytelling as a source of knowledge, and to re-center the voice of people living in LTC as a method of improving the quality of life in LTC.

The quality of care and the performance of LTC and its health care leaders can often be resolved down to a reported number to federal bodies. Empirical data can be seen as a safer, more reliable choice. However, numbers can mask or overlook the diversity of experience of the individuals that compose the data. For example, 62% of the people living in Canadian care homes are PLWD and 1 in 5 people living in LTC are prescribed antipsychotic medication, without the diagnosis of psychosis [2]. These numbers call into question the use of medication in care homes specifically in relation to PLWD and are causing national advocacy organizations to engage and question the prevalence of perusing non-pharmacological, personalized alternatives [3].

LTC Health Care Leaders at VCH had long sought new approaches to address the systemic challenges that numbers alone simply were not resolving. As Federal bodies began to recognize an increasing practice of prescribing antipsychotics in LTC, these resilient Health Care Leaders finally had the go-ahead to approach this complex issue from a new direction in their partnership with the Health Design Lab at Emily Carr University of Art + Design.

2 Partnering to Integrate Participatory Design Research Practice into Care Innovation

Health innovation is largely rooted in empirical data born out of traditional research methods. Incremental, data driven change has, to-date not facilitated the re-imaging of LTC in British Columbia as data histories offer little space innovation and transformation [4]. Workarounds to systemic problems in the delivery of care demonstrate the creativity of care staff and their commitment to deliver the best care possible despite systemic challenges. Without creating opportunities for new, holistic measures of good care that go beyond clinical need, transformation projects are forced to innovate in a system that values and evaluates using metrics that relate to past practice and struggle to understand the complexity or root of the issue at hand [5]. Operational hierarchies, for example, and task-based cultures in Canadian hospitals ensure that patients are treated as efficiently and safely as possible. Care homes however, are not the same as hospitals—they are places where people live through to the end of life, they are homes—or at least they should be. Task-based cultures, encouraged by the performance-based metrics of evaluation, can leave people in care homes feeling like they are a burden; like their life is secondary to the operations of the home. Sometimes higher-functioning people may hesitate to ask for support given the working capacity of staff and in lieu of the needs of people requiring complex care such as people living with advanced Dementia. In this way people living in care homes are triaging their needs to support care staff in performing in a way that they health system rewards.

"Quality Improvement" (QI) is the process by which most health innovation in care occurs. QI is an iterative process which takes generalizable scientific evidence, applies it to a specific context where in the impact of the change is measured, change plans are created and said plans are implemented [6]. QI offers the reliability and security in making an argument supported by research and offering confidence to leaders in making changes that can directly impact people's health. QI systems fit well within the models of education and care associated with health practice and delivery and offer Health Care Administrators a sense of security bound in history. QI, however, is thus reliant on past data, which constrains innovation to patterns of the past—not offering space for the transformational restructuring of cultural practice [4].

Participatory design methods start with the person or peoples that are the recipient of a product or service. A participatory approach facilitates the user in creating new experience-based data through meaningful co-creative engagement. Instead of looking to past research to describe a next step or support a decision, participatory strategies move with the evolutions of human connection, offering transformational possibility in actioning emergent data [7].

Living and working in an incredibly structured environment, design methods create opportunities in LTC for more divergence of thought—opening new pathways to share and understand. While difficult to test and offer the reassurances the QI system promises, emergent experience-based data is unquestionably valid. Next steps in large scale participatory projects are directed by the synthesis of volumes of diverse human experiences, documented through co-creative practice. The introduction of designers and designerly ways of doing into a health context offers space for person-directed care by stepping outside of linear solution focused systems and into the complexity connected to inclusion [5].

In 2018, Health Care Leaders at VCH embarked on a journey to shift the culture of care in LTC in partnership with the Health Design Lab and its team of Graduate and Undergraduate Design Research Assistants. Driven by an initial desire to reduce the prescription of antipsychotics, the integration of participatory design methods allowed Health Care Leaders to better understand the needs and perspectives of people living in care. This, the Culture of Care project, seeks to create a sustainable change in LTC, in co-designing a new culture led by people living in LTC.

3 Culture Creating Strategy

Harvard Business magazine defined strategy as "… a set of guiding principles that, when communicated and adopted in the organization, generates a desired pattern of decision making. A strategy is therefore about how people throughout the organization should make decisions and allocate resources in order to accomplish key objectives. A good strategy provides a clear roadmap, consisting of a set of guiding principles or rules, that defines the actions people in the business should take (and not take) and the things they should prioritize (and not prioritize) to achieve desired goals" [8]. Strategy has historically been a management level task of establishing direction and clarity for people in service/production positions to work collaboratively and uniformly towards.

In the Culture of Care project, strategy begins with bringing to light the rich stories, experiences and aspirations of people living in long term care. Through co-creative workshops, people in care homes were reconnected with their personal histories, current lived experiences and their desires for their future. It is the synthesis of these rich narratives and insights called to light by the people living in long term care that iteratively shape the VCH LTC strategy for culture change. The goal of this project is to have the people living within the walls of LTC create the principles, roadmaps and rules that guide the decision making that creates their world and its culture.

In creating workshops, media, communication materials and toolkits HDL researchers drew from research in the fields of placemaking, sensory engagement, storytelling, mindfulness and play in order to draw out the tacit and embodied memories and ideas from a participating group of people that live with a range of visible and invisible disabilities.

Placemaking. A sense of place comprises the ways in which people form connections with physical and social space. It is the tactile, emotional, relational and symbolic bonds that are experienced within physical space and community [9]. By focusing on the social and sensory experiences of a space it is possible to build a sustained *sense of meaning* connected to a location. Activities explore how people living in care would like to engage in their community and with the physical space and systemic structures of their care homes. Questions called participants to express how they felt a sense of belonging and what creates a sense of home for them.

Sensory Engagement. Employing the senses encourages rich and meaningful story telling while supporting people with invisible and visible disabilities, namely, PLWD, [10] who represent a significant portion of the LTC community. Sensory stimuli bring out latent and surface embodied information, calling forward a deeper level of knowing and often more meaningful and personalized expressions of self [11]. Creating opportunities for people in care homes to voice latent or tacit knowledge serves to validate that knowledge and can bring a sense of value and purpose to people living in care homes.

Storytelling. is a practice used across many cultures as a mode of sharing histories and passing along knowledge. Storytelling promotes communication, expression and creates an opportunity to learn about each other. Making this form of communication a normative practice in care homes builds a powerful tool for people living in homes and care teams alike to validate and cultivate the voices and choices of people living in the care homes.

Mindfulness. Creating a practice in generating awareness helps people to move beyond habitual internal dialogues. This practice can give rise to opportunities for people to reconnect with the foundational values and feelings that give rise to a meaningful existence. The goal of designed activities aimed to generate this awareness by offering sensory stimuli and working to connect these stimuli to stories and feelings; to encourage multi-sensory manners of engaging with the world that allow a person to become more comfortable with the present. Activities call a person to pay attention to and make sense of how their body reacts to various experiences and in turn how these experiences inform their preferences.

Play. The theme of playfulness was embedded throughout the event to remove the participants from the task-oriented methods of engaging with their environments as they would in an otherwise institutional setting. Play connects readily to different time periods of life, to different forms of interaction and often calls in social elements making play and access point into exploring a person's experience [12].

4 Scaling Participation

This project and partnership is currently in its third year of collaboration, working in a flexible, iterative and emergent way towards the goal of culture change. To date this partnership has included several key phases described in the sections to follow

1. Developing themes
2. Designing communication to scale
3. Regional launch: Scaling participation
4. Tool design for cultural development.

4.1 Problem Finding: Developing Themes

The Culture of Care project began in the Fall of 2018 with initial meetings and workshops between HDL designer researchers and VCH Health Care Leaders and staff to gain an initial understanding of the current culture and challenges at hand. This included a review of previous meeting minutes and initiatives through the VCH Dementia Care Team over the past couple of years, facilitation of a participatory design workshop with the Residential Practice Team, which included activities such as mapping the flows of communication and identifying barriers to culture change, as well as a workshop with Care Home Directors to identify areas of interest for change within their homes. During this preliminary phase it became clear that people living within the homes must be involved for this cultural shift to occur. Starting in January 2019, the key objective was to gain an understanding of the lived experience in care and to have people living in LTC define what 'good care' means to them. Working in partnership with the VCH Dementia Care Team, Design Researchers with HDL visited six care homes and talked with forty-six people, asking two key questions, 'What experiences are important to you?' and 'What does good care look like to you?'. These questions were asked through participatory design workshops hosted at six care homes, following an open invitation to all care homes to participate in the project by hosting a workshop at their care home. Once a care home agreed to host a workshop, they invited the people living in the home to participate in the workshop, and their family members were also invited to participate if they desired. The workshops were co-led by design researchers and staff from the VCH Team and used image cards as prompts for dialogue around these two questions. Questions were intentionally selected to be future-oriented, to create a safe environment for people living in the home to provide input and ideas.

Thematic synthesis of thoughts, stories and feelings expressed by people living in care, were distilled into five themes through affinity mapping of detailed workshop notes collaboratively with the VCH Administrators. These initial sessions introduced these health leaders, who tend towards a more linear, task based or hierarchical approach, to analysis of qualitative data and designerly approaches to mapping complexity. This foundational basis of shared understanding of human centered, participatory research methods have built design capacity within the VCH Dementia Care Team, offering new tools and methods to reconceptualize the complex healthcare challenges of this institutionalized system.

The findings of this initial phase of research created the grounding for a culturally driven strategic change, defined by the following themes described by the people living in long term care:

Building relationships and conversations. Building strong relationships and making space for conversations that are meaningful allows people to feel at ease and find a sense of belonging within the home.

Strong relationships can support patience and empathy in helping people to get to know each other's likes and dislikes. People said they prefer to know and have a strong relationship with their care as this can often help people feel more comfortable expressing what they need and want. Similarly, friendships can help those who feel uncomfortable asking for help.

Looking after Individual Needs and Care. Each person has a different background and a different set of needs and wants. Language, food, hobbies, fitness, and access to information are all areas that can require individual care routines.

People want care that is focused both on each individuals' physical as well as emotional wellbeing. Being able to access activities that meet individual needs and help fulfill personal dreams is so important for an individual's personal growth.

Giving Time for Flexible and Spontaneous Care. Flexibility in care gives people living in care homes a sense of freedom and independence. It lets people receive care and live in care homes on their own terms, like they would within their own homes. Because care homes operate within a strict scheduling system, including meals, sleeping and social time, impromptu actions are limited as is the staff's ability to respond flexibly. By limiting flexible and spontaneous care, the current culture of care also limits voice and choice Staff may avoid creating opportunities to try new things, if they think a person may not be able to do it. Assumptions around ability, therefore, can also limit flexibility.

Recognizing and Supporting Ability. Assumptions about the abilities of people who live in care homes sometimes cause them to be excluded and can cause people to question their own ability.

Sometimes, in an attempt to provide care, a culture can be created that limits how people think about their own abilities—holding people back rather than encouraging their independence.

Creating a Sense of Purpose. Before living in care homes, people's activities are generally purpose-driven: cooking for family, walking to the store to get groceries, building skills to reach personal goals. But after moving into care homes that sense of purpose is often lost, since all of a person's basic needs are cared for. Many people are rewarded by and expressed a desire to help and support their home and community or participate in activities that offer fulfillment instead of entertainment.

People living in homes can begin to see themself as patients, rather than as individuals with free will and aspirations.

4.2 Scaling the Reach of the Voices We Heard: Communication Designed to Re-Frame the Future

Part of shifting culture is offering a new way of looking in order to create the space for seeing the world in a new light. In choosing a method to share the five core tenants of the new culture of care, the HDL Research team was especially attentive to the manner in which an aging population is spoken to, and the manner in which they are portrayed. By reconsidering, reframing and trying different methods to shift the traditional ways people living in care homes are communicated to while shifting the modes of communication received from and portrayal of people living in care homes. Researchers did this by developing evocative visual systems that were carried through within all forms of communication sharing. Content created would be shared with all 55 long term care homes by VCH staff and needed to service the unique needs and audiences at each home.

The HDL team elected to create an animated video, narrated by people living in care homes, with a companion PowerPoint slide deck that would allow facilitators from VCH to tailor content and questions to the unique needs of their home, allowing for continued conversation (Fig. 1).

Vibrant colours and collage inspired imagery painted people in care in a new light; inspired by the vibrant clothing and spirit of the video's narrators, each slide and storyboard were created to emphasize the wealth of personal narrative and diverse identities within the homes. In depicting the stories, the team elected not to use people's faces but rather used birds and people's hands as symbols of new beginnings, global connections, and a collective diversity while maintaining a unique identity. Shapes were used to describe communication and language difference offering a more inclusive way of describing difference. Videos included text captions in both English and Cantonese for those hard of hearing and were posted to YouTube in hopes that translation services could support those with different language needs.

In order to test and prototyping ways to express themes, the HDL team set up Pop-Up recording stations at select care homes and recruited people living in homes to bring the previously collected quotes to life, in voice acting their stories for use in the subsequent animations. Arriving into the care home with a laptop and a microphone proved that flexible and spontaneous activity was within the staff's capacity. Participants while often feeling initially that they may not be able to read or recite effectively, ultimately felt a sense of accomplishment; of their voice having meaning; that their ability was being supported. The creation of the videos, thus, served as a proof of concept, prototyping the effects of implementing the themes.

The inclusion of VCH care staff and Dementia Care team throughout the process of production of the video's companion slide deck offered a nuanced language which could engage both the care staff and people allowing for a multi-

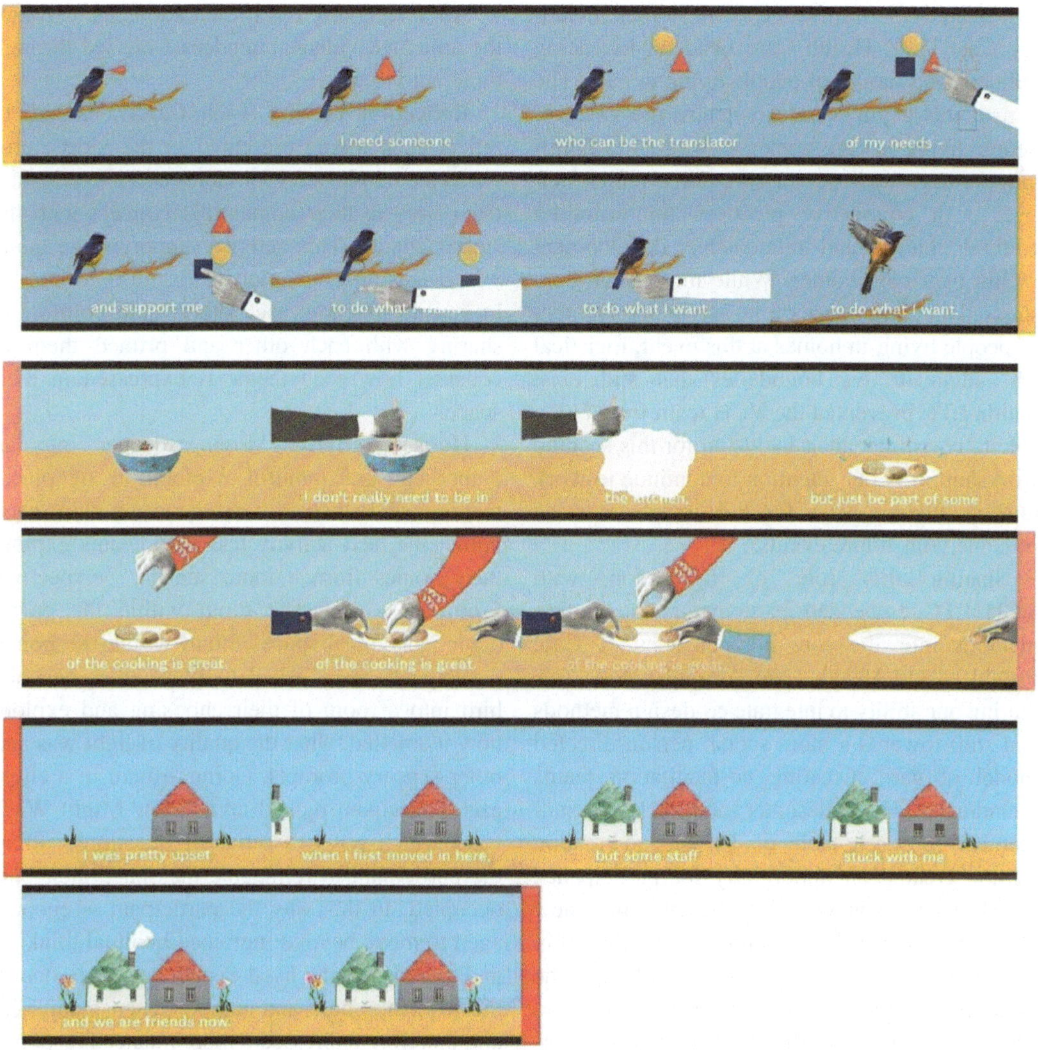

Fig. 1 Stills of videos which illustrate person-developed themes can be found online in both English and cantonese and are being used to share the core tenants of the culture of care within care homes and beyond into the community

dimensional approach to how the five themes might be workshopped. Collaborating in this way positioned the materials as flexible and adaptable, allowing staff to consider the relevancy and specific messaging that could best support their home; to see themselves within the content. Creating opportunities for all those in contact with the content to be a contributor that hierarchies and ingrained manners of working have space to shift.

4.3 Regional Strategy Launch "Come Alive": Create the Space for Health Care Workers to Lead Co-creation Activities

In October of 2019, VCH LTC, in partnership with HDL, launched the Culture of Care project to the health region in a 100 + person event called "Come Alive", which was attended by representatives from sixteen care homes including Care Staff, Health Care Leaders, Licencing Officers, and family of people living in care. The aim of this event was to re-centre the voice of people living in care homes as well as to familiarize a large audience of care home representatives with co-creative practices for visioning cultural change and relationship development within their care homes. While the design team stressed the importance of the active involvement of people living in homes at this event, logistical and administrative boundaries that still exist within LTC prevented the VCH team from being able to coordinate their inclusion for this specific workshop. It is our intention to continue to work towards finding ways for this inclusion to be possible with future events.

Sharing this role of facilitation with VCH LTC embedded participatory facilitation methods into the core competencies of Care Providers and Health Care Administrators, furthering our ability to integrate co-design methods and shift towards a more social, person directed model of care. Creating co-facilitation teams including a HDL designer and a VCH staff member for each table of 10; the co-creation session created an opportunity for participants, largely staff, to view their colleagues in a new light and see a shift in approach coming from within. Stories of care were shared while at the same time attendees were able to listen and support each other in the challenges of the current institutionalized system.

Upon entering the event attendees were given a nametag and a coloured bird which they were encouraged to decorate at a table of their choosing with a range of available materials. This craft-based activity engaged attendees' bodies, taking them away from their phones and opening up space for dialog and creativity. A key learning from this event was that tangible, co-creative activity for those engaged in the work of health supports a richer more engaged dialog as these individual's normative practice to multitask (Fig. 2).

After receiving the presentation, inclusive of the animated videos, attendees were led through three activities:

Reflection Activity. Table Cards, designed to reflect each theme presented in the video, provided a tangible asset for tables of 12 to consider the voices in their homes that connect with the stories shared. This activity supported the table, which had representation from multiple homes, in getting to know each other, gain comfort in sharing with each other and primed them to consider how each theme is expressed in their space.

House Activity. Moving deeper into an embodied and mindful exploration of place-making and thematic resonance within each home; the next activity had participants explore their homes from a more sensory perspective. Prompted to select a room within the poster depicting typical spaces found within the home, participants were asked to place their bespoke bird into a room of their choosing and explore how it smelled, what the quality of light was and other sensory prompts. Going around in a circle each participant picked up a "How Might We?" Card which posed a series of questions calling the participant to reimagine the space their bird occupied. In this way the participant is encouraged to move beyond ingrained habitual thinking and to consider the lived experience. This step in the workshop built and embodied sense empathy and therefore expressed a more nuanced response to possibilities within the space beyond top-of-mind ideas.

Tree Activity. Finally, after having considered the possibility of a new experience within the walls of the home, participants were guided through a visioning process that called them to consider from a social, cultural and emotional perspective where they would like to see their home move towards with corresponding questions that pushed participants to develop blue sky type vision for what culture change could look like.

Fig. 2 Co-facilitation building familiarity with the delivery of co-creative workshops in healthcare workers

4.4 Tools Designed for Person-Directed Culture Development

In the months that followed "Come Alive", the VCH Dementia Care team, having built confidence and proficiency at leading participatory sessions, facilitated both the presentation and select activities from the October 2019 event to all fifty-five long term care homes in order to onboard the region to the work to come.

Upon hearing the videos, many participants recognized their own voice and there were requests from people in homes to get the opportunity to share their own stories. Oftentimes, the tools developed were used in an adaptive manner in order to serve the specific needs of each care home. The message, tone and vibrancy of the videos resonated with people living in homes and in-turn Health Care Administrators and care teams saw how these messages applied to the culture that they too work within. Residents did express apprehension, however, about exactly what was going to happen and when. Some residents expressed a clear call to action; that their time was limited and that an urgency and commitment to change was important.

As the Dementia Care team shared the presentation slide deck and co-creative activities with the broader region, the HDL team set to work considering how to begin to further explore the five themes presented in a participatory manner at all fifty-five VCH long-term care homes. As a first step to shift institutional hierarchies and re-imagine the unique culture within each VCH care home, the HDL researchers considered approaches to activities that could engage staff, people and community in developing personalised visions describing possibilities for contribution to care.

An important consideration in the design of the next step of the project was how to scale participation. How, could a team of four designers co-create with an entire region? Having co-facilitated within the VCH team in the space there was a small capacity for those employed by the health region to support—however it was important that people living in homes themself become directors of their own life in ways that felt important to them. Thus, the team selected the form of a game, one that could be easily reproduced and live contextually within shared space in the homes.

"Time Traveler" Game Theory Inspired by the Life Café [13] toolkit designed by the Lab4Living, "Time Traveler" is an activity that brings people together and provides prompts and provocations for storytelling that supports the development of relationships and conversations. The game facilitates reflection upon when and how people have felt purposeful in their pasts, thus imagining what that Sense of Purpose might look like in the care home in their present. The relationships built through creating a practice of storytelling and listening serves as a means of spontaneous support in an otherwise highly structured system—it proves that care can be found in the untapped resources of the people living in care, themselves in addition to other community members. Storytelling also makes space for people to express personal histories of accomplishment, hardship, value and meaning—in sharing these stories there is space created for others to see and seek out these traits and values from the individual allowing for the care home to more organically and authentically recognize and support the abilities of people.

Finally, a goal of this activity is for people to feel more self-expressed, for their individuality, for requesting or creating the conditions that serve their holistic needs and care in sharing and creating a new and personalized narrative built off the past and re-centered into the present.

Objects used in the game were selected based on analysis into their power to surface strong feelings and memories from those who interact with them. In sourcing and understanding objects to include in Time Traveller, researchers looked for objects with an uncanniness, relationality and transitionality [14]. Consideration in object research spanned cultural association, key global events, personal transitions and rites of passage, and moved from universality to cultural specificities for the first iteration of list of objects considered for time travel. Finally, the team did a sensory survey of objects in this investigation into their evocation.

When considering the form of time travel the researchers focused on the idea of familiar ritual, awe and gift giving. The desire was to create a sense of embarking on a new journey, on learning and one sharing. Thus, having time capsules in the form of small boxes to be easily opened and revealed became central to our journey through time.

While the evocation of objects brings to the fore a richness of personal narratives and hints as to personal aspiration or value, the sense of time travel breaks game players out of a task based or linear system deeply ingrained within the lives of staff and thereby people living in homes. This disconnect from the linear borne out of play creates space for ideation within the home that may give rise to new opportunities yet unexpressed.

Co-Development with VCH LTC In a series of co-development sessions, the researchers worked within care homes across the VCH region to understand connection, interaction, relationships created between objects, and individuals.

These in-context interactions offer designers a view into health systems, caregiving and living within the walls of a care home. Imbedding design research assistants within the health space builds the capacity of this trade to better support health systems in future. While the students outside and unbiased perspectives on possibilities for culture change can create the space for new ideas—a grounded understanding of the lived experience and in person interaction with the people that live in care ground ideas in reality. Trialing the game and understanding capacity of staff supported the designers in better understanding the challenges designing for visible and invisible disability, breadth of cultural diversity and connection of the game's intention with its participants.

The in-home sessions supported the development of a guide which could be easily administered either by a person living in care or care staff member. Co-development supported and informed the design of the game in the space of participation, and in how people in a broad range of mental and physical health might be able to participate while attuning the researchers to the rich cultural diversity that exists in LTC in the Vancouver region.

Game Design for Scalable Participation: Time Traveler The Time Traveler takes its participants on a journey back in time to recall old memories, share experience and reflect on how those experiences and sentiments can be valuable in the present. It uses iconic and tactile objects to trigger memories and concurrent emotional associations through sensory provocation.

The activity begins by unveiling a suitcase that contains tools for a journey into a person's past. These tools include:

- Time capsules or a series of 9 covered boxes containing evocative objects
  - Evocative objects that operate as tickets to travel and as memory triggers.
- Feeling Cards to describe and support a felt description of memories
- Postcards that act as souvenirs and qualitative data collection tools
- Facilitation guide which operates as the conductor to time travel ensuring transportive experience and a rich exchange.

To embark on this journey to the past, the facilitator asks the participants to select an object that they are drawn to, then probes for its connection to story and to the past. The participants follow each other along into their personal histories learning about the unique narrations of each individual's life. Participants are asked to deepen these stories by selecting a feeling card that is associated with their story and describe how this feeling is linked to their memory. The participants are then asked to carry that feeling they connected back into their present and to consider if they have the same emotional associations that surface. When do they feel the same emotion, feeling, sensation, interaction etc. in their care home? This initiates a moment of being present, connected to the past, a moment to recognize their own voice, the relationships being fostered around the table and the engagement participants have with their care home and community (Fig. 3).

Personal observations can be scribed onto postcards allowing each home to slowly build a picture of the stories, values and personal histories of the people within the home.

In March of 2020 As the team prepared to produce and deliver 55 Time Traveller toolkits at and 100 + person event that included residents from across the health region the COVID 19 Pandemic began to emerge and called all personal and tangible interaction with populations in LTC to a halt.

5 Pandemic Pivot

The COVID-19 pandemic has called for a pivot in approach in the researchers' interactions with long term care. In efforts to keep people in homes healthy and safe strict protocols have been implemented often limiting opportunities for connection and culture development. Meanwhile, families expressed their heartfelt desire to communicate with and support their loved ones through this challenging time. With the health system focused on saving lives, and protecting staff, the HDL research team has started to engage families in co-creative virtual workshops

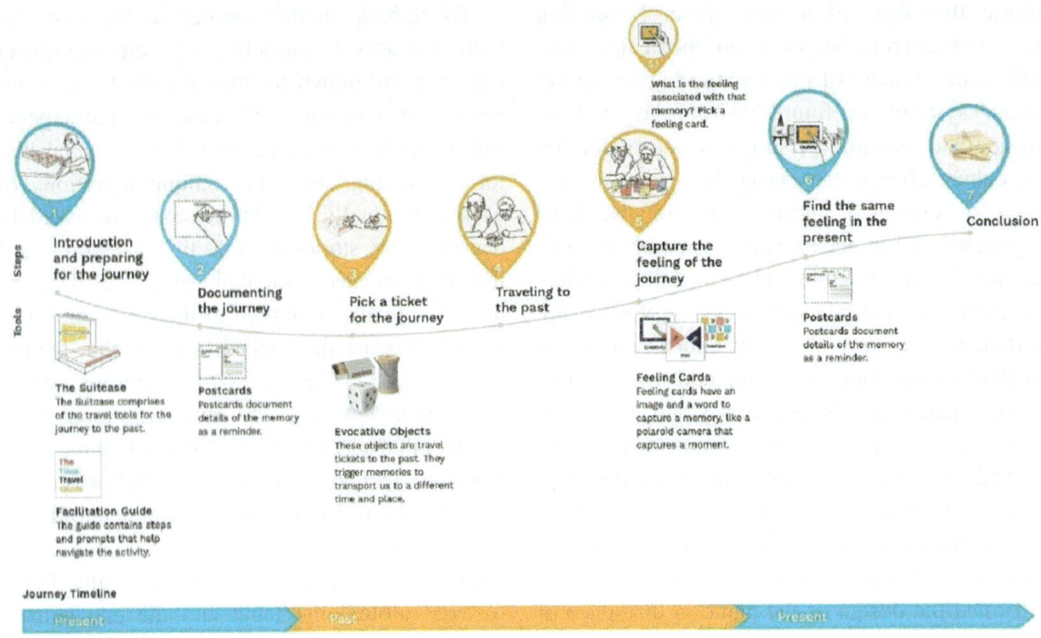

Fig. 3 Time traveler activity flow

in order to ideate possible new modes of connection. This difficult moment has presented a new opportunity to extend the culture within the home, out, and to weave family more intentionally into the culture of care within long term care.

6 Conclusion

Shifting to a person driven culture within LTC is not simple and it is not fast. LTC culture has been bound in assumptions around ability and ingrained in a rigid working culture. In collaboration with designers, LTC can gain proficiency in divergent thinking and comfort in emergent change. Designers can facilitate this transformation in providing approaches and strategies that center and elevate marginalized voices. In visualizing complexity, designers offer the security and active participation that can support Health Care Administrators in moving beyond data driven decision making.

Finally, in supporting communication and collaboration across all stakeholders the design research process collapses health silos creating the opportunity for communities to re-create themselves led by the voice and choice of the people who live the system: “It’s those voices that will inform and propel these traditional institutions into vibrant places where people thrive and to come to life, where they are heard and are treated in a way we should all be treated.”—VCH Health Leader.

References

1. Brooker, D.: What is person-centred care in dementia? Rev. Clin. Gerontol., 215–222 (2003)
2. CIHI.: Potentially inappropriate use of antipsychotics in long-term care (Percentage), 2018–2019 (2020). Retrieved from Canadian institute for health information: https://yourhealthsystem.cihi.ca/hsp/inbrief.#!/indicators/008/potentially-inappropriate-use-of-antipsychotics-in-long-term-care/;mapC1;mapLevel2;/
3. Alzheimer Society Canada.: Alzheimer Society Canada (2017, June 7). Retrieved from use of antipsychotic medications to treat people living with dementia in long-term care homes: https://alzheimer.ca/sites/default/files/files/national/media-centre/asc_position_06072017_antipsychoticmeds_e.pdf
4. Molloy, S.J.: Innovation labs in healthcare. Toronto, OCAD University (2018)

5. Anna Meroni, D.S.: Massive codesign: a proposal for a collaborative design framework. Milano, Open Access (2018)
6. Batalden, P.B., Davidoff, F.: What is "quality improvement" and how can it transform healthcare?. Qual. & Saf. Health Care **16**(1), 2–3 (2007). https://doi.org/10.1136/qshc.2006.022046
7. Brown, A.M.: Emergent strategy. Chico, AK Press (2017)
8. Watkins, M.D.: Harvard Business Review (2007, September 7). Retrieved from Demystifying strategy: the what, who, how, and why: https://hbr.org/2007/09/demystifying-strategy-the-what#: ~ :text=Here's%20my%20definition%3A%20A%20business,desired%20pattern%20of%20decisi
9. Ensiyeh Ghavampour, B.V.: Revisiting the "model of place": a comparative study of placemaking and sustainability. Urban Plan., 196–206 (2019)
10. Alba Sánchez, J.C.-C.-L.: Multisensory stimulation for people with dementia: a review of the literature. Am. J. Alzheimer's Dis. & Other Dement., 7–14 (2013)
11. Heersmink, R.: The narrative self, distributed memory, and evocative objects. Philos. Stud., 1829–1849 (2018)
12. Ayse Sentürer, S.S.: A view to space and design through play: a research project at the architectural design studio. Int. J. Arch., Spat., Environ. Des., 41–51 (2016)
13. Helen Fisher, C.C.: Life cafe. Lab4Living. (n.d.). on %20making.&text=Together%2C%20the%20mission%2C%20network%2C,strategic%20direction%20for%20a%20business
14. Turkle, S.: Evocative objects—things we think with. Cambridge, MIT Press (2011)

The Challenges of Creating Design Requirements for Products for People with Dementia

Rita Maldonado Branco, Niels Hendriks, Lieke Lenaerts, and Andrea Wilkinson

Abstract

With an ever-growing number of people with dementia, no cure, treatment or effective prevention effort is available to bring medical relief to deal with this condition. Therefore, efforts should be made to support people to live well with dementia and design can contribute to this by the creation of better care services, products and environments. A European research and implementation project was set up to tackle this, with the aim to promote and support the development and quality of products designed to enhance the daily lives and care of people with dementia, including the creation of a certification quality mark. In order to set up that certification quality mark, a series of requirements for 'good' design for people with dementia need to be developed. This has presented several challenges due to the diversity and complexity inherent to dementia, as well as to the wide range of products on offer. This paper presents the initial phase of this ongoing research project and discusses the uncertainties and critique regarding the creation of design requirements and how the consortium overcame them.

Keywords

Design requirements · Dementia · Certification

1 Introduction

In recent years, pharmaceutical research focusing on finding a treatment for people with dementia has not provided a lot of hope that a cure would be found in the near future. In 2018, two pharmaceutical companies working in this area announced that they were ending their research into finding a drug that could cure or prevent dementia [1, 2]. In addition, many prevention efforts are limited to actions that are generic such as maintaining a healthy lifestyle, avoiding obesity, diabetes and smoking, emphasizing exercise and the engagement of social interaction with others [3]. At the moment, then, there is no medical solution that will bring relief to the ever-growing numbers of people with dementia. This results in a situation in which the response from design includes the creation of better care services and better products (and environments) to support people with dementia in their daily life and care.

R. M. Branco (✉) · N. Hendriks · L. Lenaerts · A. Wilkinson
Research Unit Inter-Actions & Dementia Lab, LUCA School of Arts, Genk, Belgium
e-mail: rita.maldonadobranco@luca-arts.be

R. Brankaert et al. (eds.), *Dementia Lab 2021: Supporting Ability Through Design*, Design For Inclusion 2, https://doi.org/10.1007/978-3-030-70293-9_2

This paper highlights the first months of the *Interreg North-West Europe Certification-D* project. The goal of this European research and implementation project is twofold. On the one hand, it wants to provide a framework that enables companies to design better products for people with dementia. On the other hand, it wishes to support people with dementia and their network of family and caregivers in selecting products that will be suitable for them or their loved ones with dementia, through the creation of a certification quality mark for products for people with dementia. The project includes a variety of partners (care, advocacy groups, academia and industry) and comes from five different European countries. The products being examined deal with three categories: 'safety and security', 'enabling' and 'leisure time'. Four Living Labs involving people with dementia were created to support product evaluation and the development of the certification process, each focusing on one of the three mentioned categories. The authors of this paper are part of one of the Living Labs, which deals with leisure products. The greater goal of Certification-D is to enable people with dementia to live longer and in a qualitative way at home and thus, it includes a strong focus on quality of life.

The development of a certification procedure and accompanying certification quality mark includes developing requirements that propose what a good design for a person with dementia can be. In this paper we focus on the research that was used to create these requirements. This list of requirements will serve as a basis for evaluating specific products and to check whether a quality mark can be granted (a mark that would indicate that the product is made in a dementia friendly way), as well as a set of considerations for designers and manufacturers who aim to make products for people with dementia.

In the following sections, the paper will discuss the challenges and uncertainties that arose in the first research phases, related to the complexity of creating a list of requirements to evaluate products for people with dementia. The paper will give an overview on how the consortium tried to overcome these challenges, through an expert-led study including a preliminary workshop with consortium members, a Delphi Study with experts outside the consortium, and the already mentioned setup of four Living Labs. It is important to indicate that the first research acts, the preliminary workshop and the Delphi study, did not directly involve people with dementia. Despite the authors' extensive experience and sincere belief in the involvement of people with dementia in every research phase, the Delphi study only involves people who do not have dementia. In our view, the Delphi study is no more than the necessary run-up to the participatory phases (the Living Labs) where the involvement of people with dementia is key. We are confident that even without involvement in these initial phases of the research project, we are not ignoring the voice of people with dementia. Moreover, we believe that the way the project was set up puts people with dementia central, meaning that, from the very beginning of the project, those who advocate for people with dementia were involved in defining how the project should be set up and come about. As Hendriks et al. [4] indicates, having a care organization as a primary participant in research projects, instead of as secondary or contextual partners, is important in the organization of design research projects. Without this specific inclusion of care partners and organizations as full partners, projects such as this run the risk of creating a power imbalance, in which a 'cult of experience' exists. Academic partners and partners from industry (who often take the lead in design projects) become positioned as experts and 'lay' partners, in this case coming from care, can be less experienced in design and research projects. This immediately created a working position in which the nuances, challenges and complexity of working with and for people with dementia was prioritized.

2 Challenges & Uncertainties About Creating a List of Requirements for 'Good' Products for People with Dementia

The basic idea of creating a list of requirements to evaluate products for people with dementia was initially met with criticism. The authors of this paper were, from the start, critical towards the overly ambitious goal that *Certification-D* set forth. There was little belief that a one-size-fits-all series of requirements could be used to evaluate each and every product. Their awareness of the condition of dementia added to this critique. In addition, they were concerned that a list of requirements could be understood as a recipe that can be applied too rigorously in the hope of leading to a satisfying outcome, without the need to take into account the situation or context of use.

Specifically for this project, the context of use involving people with dementia, makes the creation of a list of requirements difficult. Dementia affects each person very individually: one person might have difficulties speaking or remembering words, while others won't; one might have to deal with impaired motor skills while another might hardly be affected by this. Besides being a very personal condition, dementia also evolves over time, resulting that a certain ability might be lost after a couple of weeks or months. Apart from the condition of dementia, the idea of evaluation as a tool to define whether a design is well or ill-suited, also includes critique [5]. Different standards and expectations of each individual make it hard to create one tool for evaluation that would be suitable for them all. It is thus correct to say that the unpredictability of a person's wants and needs, and the way dementia affects someone makes certification a challenging task.

The amount of research documenting similar efforts in finding a type of standardization or set of rules to design for people with dementia is limited. The remainder of this section presents a summarized overview of a literature study, looking at research regarding design requirements, guidelines, recommendations and design considerations, and including references from academia and (inter)national dementia associations and networks, such as the Alzheimer's Society or the Dementia Engagement and Empowerment Project (DEEP).

The endeavors to create lists of requirements mostly came from the research and process of designing a specific product or group of products. These included, on the one hand, specific criteria based on primary research and/or on literature, to be addressed by the product that was to be designed (e.g. [6–8]); and on the other hand, learnings yielded from the experience of designing a product and often testing it with people with dementia. These learnings are usually shared in the form of generalized guidelines or recommendations for similar types of products (e.g. [6, 9–13]). Other approaches from research sought to define general recommendations and were put together as principles or criteria for successful designs of a specific kind (e.g. technology-based products, environmental design, etc.). These were often based both on literature and on a broader body of experience from different projects developed in that field (e.g. [14–17]); or they were framed as more general considerations, focusing not only on how to design for people with dementia but also on how to codesign with them (e.g. [18–24]). Dementia associations and networks often presented more practical, short and direct bullet-point guidelines (e.g. the DEEP guides [25]) regarding specific topics such as language or dementia-friendly websites.

This literature review suggests that, in some cases, it might be possible to establish a precise set of guidelines, namely for specific types of products or regarding particular aspects of products, such as the Dementia Digital Design Guidelines [26], or the recommendations on sensory rooms for care homes made by Jakob and Collier [16]. However, it seems difficult to create requirements that encompass a wide range of products. Although the literature study identified research that resulted in guidelines and criteria, no published findings could be found that had a similar goal as *Certification-D*: to bring together a series of requirements that would define a good design (regardless of the product being a digital or physical one; being used in group or individually; etc.) and that would go beyond general recommendations. This made us consider whether creating such a list would even be possible.

In short, the basic premise of *Certification-D* was met with some criticism at first: ending up with a list of requirements was perceived as highly challenging due to the wide range of products involved, the diversity of the dementia condition and the individuality of expectations and needs of each person. In addition, literature showed a difficulty in defining these types of lists, as they were mostly too product focused. Moreover, the variety of products integrated in the project made it even more important to reflect beyond one specific product category. However, as indicated above, we felt this project touched upon a necessary thing, namely, to make better products for people with dementia.

3 Tackling the Challenges and the Uncertainties

3.1 Expert-Led Study

As indicated, in order to take into account the complexity of the dementia condition, caregivers and advocacy groups were put central in the project. However, to cater for the variety of product categories, also other experts (e.g. product designers, SMEs creating digital tools for people with dementia, researchers focusing on quality of life, etc.) were added to the consortium or were asked to join at least one of the three phases of the creation of the requirements. A preliminary workshop was conducted with the consortium partners and wished to create the initial list of requirements for each product category (enabling, leisure, safety/security). These and other experts were involved in the other phases: a Delphi study contributed and validated these initial lists and the creation and use of Living Labs, which allowed for people with dementia and their family and caregivers to further validate and add to the resulting lists.

Preliminary workshop. As a leadoff for reflection, a workshop was organized in order to gather input from the *Certification-D* project partners for the development of an initial certification list, using existing products for people with dementia. In the workshop, participants were divided into three groups. Each group evaluated three different products (a total of nine products were evaluated), ranging from analogue objects like a coloring book to electronic items, such as a tracking device. After receiving a product and additional information about it, groups wrote down general and detailed requirements concerning the product. The workshop ended with a discussion facilitated by an interactive mural with post-it notes providing insights into recurring categories for each product category (Fig. 1).

Delphi study. Following the preliminary workshop, a Delphi study was set up. A Delphi study aims to collate expertise in order to generate consensus, usually over several rounds of participation [27]. For each category, a pool of experts from dementia care and dementia-related design research and practice (e.g. companies that make products for people with dementia) was created. Based on literature, on the outcomes from the preliminary workshop and on the concerns described earlier, lists of requirements were created for each of the three different product categories: leisure, enabling and safety/security. These three lists were the starting point for the experts' contributions, who were asked to add to, delete and comment on the list regarding the category they were assigned to. The Delphi study was composed of 3 rounds and was conducted over March and April 2020, with 22 experts participating in at least one of the rounds.

Living Lab evaluation. Four Living Labs (i.e. Belgium, France, Germany, and The Netherlands) have been set up to involve people with dementia [28], to validate and further refine the lists of requirements that resulted from the Delphi study. As mentioned earlier, the authors of this paper are part of one of these Living Labs,

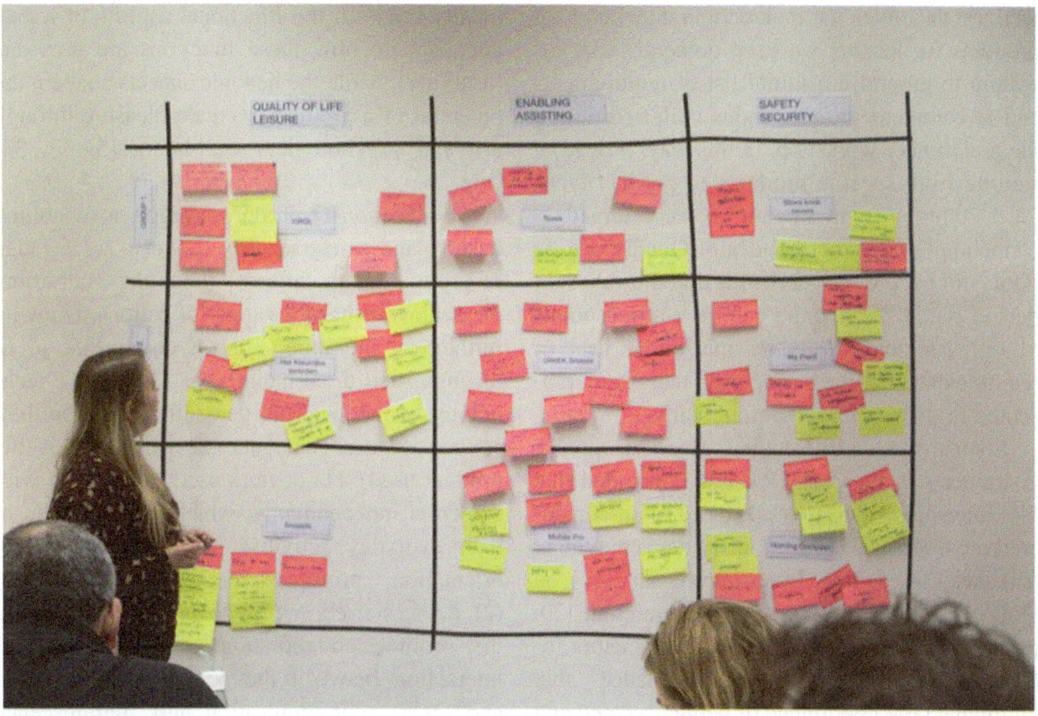

Fig. 1. Discussing requirements during the preliminary workshop

which focuses specifically on products created for leisure.

In order to evaluate the list of requirements, people with dementia and their caregivers were invited to try different products already on the market that have been deemed as 'good products' (i.e., tested and successfully used together with people with dementia). By evaluating 'good products', the intention was to identify potential flaws in the lists of requirements. The evaluation of these products took place in the real-life environment of the participant, meaning the homes of the people with dementia or a day care center where they reside. In the Living Lab that the authors of this paper are responsible for, the participants used and evaluated a product for two weeks and were asked to answer product-specific question-cards. The first time the participants used the product was observed by the researcher. After the two weeks, the researcher went back to the participants for the reflection phase; an open interview guided by the answers to the question-cards. The aim of the interview was to understand whether requirements were missed and to identify necessary changes that could be made to the list of requirements.

3.2 Creating the Initial Lists of Requirements: Person-Centered Dementia Care and Quality of Life as a Starting Point

The ultimate aim of *Certification-D* is the provision of quality of life for people with dementia, namely those still living at home. Through their own experience as researchers and designers, the authors were aware that the context of home and the concept of quality of life creates a very complex research space. Not only does it balance between concepts of independence and identity, but the concept of home is a divergent care context. In order to tackle the challenge of the complexity inherent to the diagnosis of dementia and to what means to live well with dementia, as well as the diversity included in the range of products for leisure, we used dementia care literature to ground our initial list of requirements and to complement the insights collected during the preliminary workshop. In this way, instead of starting from specific products or product types (as is often seen in literature), we returned to Dementia Care, using the idea of Quality of Life (QoL) for people with dementia and the values of person-centered dementia care as starting points to organize and formulate the initial lists of requirements for each category that were presented to experts in the Delphi study.

Ettema et al. [29] studied dementia-specific instruments to measure QoL and concluded that the domains of QoL that are more often considered are affect, self-esteem, activities, enjoyment and social interaction. In a more recent study on QoL of people with dementia, Verloo et al. [30, pp. 2141–2143] identified four major factors for QoL: (1) "human dignity and acceptance", that highlighted the importance of being accepted and maintaining social relationships with family and friends; (2) "development and existence", concerning the importance of activities and hobbies, seeing them as "recreational moments that allow them to feel real pleasure"; (3) "functionality and health", regarding the maintenance of both physical and psychological health (which goes beyond managing their illness); and (4) "recognition and safety" that emphasized the importance having respect for everyone's privacy, life history and personality.

These perceptions on QoL are also aligned with the psychological needs of people with dementia identified by Kitwood [31, pp. 80–84], and further developed by Brooker [32, pp. 96–99], which are comfort, identity, attachment, occupation, inclusion and an overall need for love.

The ideas on QoL and psychological well-being of people with dementia suggest that not only utilitarian aspects of designed products are important but also those that go beyond functionality. Hassenzahl [5] takes an explicit stance in this by distinguishing between pragmatic and hedonic qualities of products. The pragmatic quality refers to the functional aspects of a product, and to how these functions are provided (usability), while the hedonic aspects concern the potential of a product to create pleasure through use and psychological well-being [5, 33, 34] (Fig. 2).

Certification-D product categories (enabling, leisure, and safety/security) cover a wide range of products with different functions, supporting the achievement of a variety of actions. Drawing further on literature, Gibson et al. [35, p. 6] distinguished three main types of assistive technology for people with dementia, based on their context of use (who and in what way is the product used): (1) devices used *'by'* people with dementia independently, which support them in their everyday activities (devices that support orientation, prompts, reminders, aids, etc.); (2) devices used *'with'* people with dementia, that promote collaboration, communication and interaction between the person with dementia and others (communication aids, reminiscence and leisure products); and (3) devices used *'on'* people with dementia, which do not involve the engagement of the person with dementia with the product, although used to support the person's life (such as telecare, monitoring and safety devices, etc.) (Fig. 2).

Broadly related to the categories defined for this project, this arrangement is useful to distinguish different kinds of actions and motivations for actions and contexts of use, which are relevant for defining the criteria to evaluate the different categories of products. Although pragmatic and hedonic qualities are important for all kinds of products, they might be more or less relevant, according to a specific product. For example, a device used 'on' people with dementia (e.g. gas alarm) is more likely to focus on pragmatic qualities (the alarm needs to ring when needed), while hedonic aspects like stimulation, evocation, identification and aesthetics are less relevant (although one might not want to have a gas alarm that visually interferes too much with the space where it is placed). However, this can also be defined by the situation [33]. If the product is, for instance, a GPS tracker to be

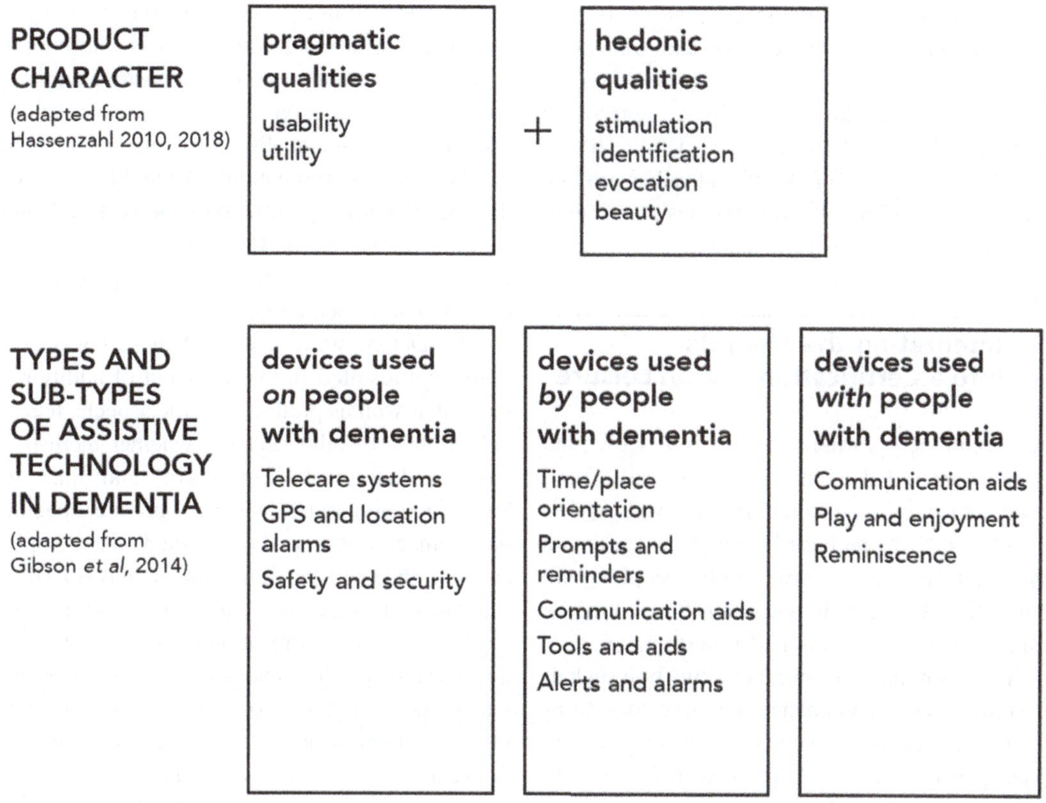

Fig. 2. Product character, adapted from Hassenzahl [5, 34]; Types and sub-types of assistive technology in dementia, adapted from Gibson et al. [35]

visibly carried by the person with dementia, then the hedonic quality plays an important role, namely the aesthetics and identification aspects.

3.3 Rethinking Design Requirements for Products for People with Dementia

For *Certification-D* the list of design requirements not only needed to deal with dementia as a condition that affects people in different and very diverse ways, but it would also need to encompass the complexity of dementia experience and include a broad group of products that differ in function (the three categories: enabling, safety/security and leisure) and in "type" (e.g. digital, electronic, analogue). In order to be used in a certification process, these requirements needed to be general enough to cover this variety as well as leaving space for the specificities of different kinds of products. To address these challenges, the initial list sought to be extensive, combining both general and product-specific *considerations* that should be taken into account when designing a product for people with dementia. Thinking about *considerations* ("something that must be thought about when you are planning or deciding something" [36]) seemed more appropriate to deal with the complexity of dementia and the diversity of products, rather than a strict and fixed list of *requirements* ("something that you need or want; something that you must have in order to do something else" [36]) or *guidelines* ("a set of rules or instructions that are given by an official organization telling you how to do something, especially something difficult" [36]). This list of considerations would serve as a basis for selecting and building a list of specific and pertinent requirements for evaluation, depending on the product category, characteristics and context of use. In addition, this list is seen as a "go-to" document that could be used by designers and manufacturers. It will offer not a definite list of criteria with boxes to tick off, but rather provide considerations that will aid in the design of better products for people with dementia.

4 Integrating the Insights into a Certification List on Leisure

As mentioned earlier, studies on QoL and person-centered dementia care, on user experience and product evaluation, as well as the insights from the preliminary workshop based on the partners' experience with people with dementia were taken into account for creating an initial list of considerations for certification.

Themes related to a product's function such as usability, to its maintenance (e.g. hygiene), to its documentation (e.g. manual of instructions and packaging), as well as to possible risks and ethical issues associated with the product use were highlighted during the workshop as prominent points for product evaluation and certification, and thus included in the initial list.

The values of person-centered dementia care—valuing people regardless of their age and ability, providing individualized care, recognizing each person's perspectives, and promoting a rich social environment [32]—were integrated into the proposed list of considerations, ensuring a deep respect and acknowledgement of the uniqueness of people with dementia across all the document, and particularly emphasized in the ethical issues; the provision of individualized care gave origin to relevant sub-themes such as 'personalization' and 'flexibility'; the psychological needs of people with dementia [31, 32] were also embedded on the theme of utility and usability, which was framed in a way to support comfort and inclusion of people with dementia, and more specifically through the inclusion of a theme on the 'quality of use', with a view to encourage enjoyment and social engagement. The approaches that focus on designing for positive experiences and well-being mentioned earlier were found to be particularly relevant, namely in the case of leisure products whose main functions are usually to promote positive engagement and occupation, and social interaction. Therefore, hedonic qualities were also included in this list, such as stimulation ("the ability of a product to provide opportunities for personal and skills development" [34]), which was also associated with the psychological needs of people with dementia described above.

The Delphi study deepened these considerations, and resulted in the inclusion of additional general as well as product-specific aspects. It also led to other insights, namely whether there is a need to distinguish 'must have' and 'nice to have' requirements, whether product-specific requirements were to be included, the compliance of this certification with already existing standards and regulations, and an overall need to work towards uniformization between the lists of the different product categories (should they be made more uniform or more specific) both regarding their structure, format and content.

Next, we present a brief overview of the final leisure list that resulted after the Delphi study. Although it is mentioned earlier that the list is composed of both general and product specific considerations and requirements, in this overview, we can only present the overarching general themes as the full list is still under embargo. The leisure list is organized into 3 main parts, a first one about the product, its description and the manufacturers' intentions for the product; a second one regarding the user experience with the product; and a third one about other important considerations, such as ethics and risk. Main themes include, in the first part (1) *context of use*, characterizing the situation where product use is intended to take place; and (2) *before use*, regarding features that are important before using the product, such as packaging and instructions on the set-up or installation of the product, as well as aspects that might be important at every time the product will be used, such as recommendations or suggestions of different activities that can be done with the product; in the second part (3) *utility and usability* refers to functional and pragmatic aspects of products, such as accessibility, simplicity and familiarity; (4) *daily use and maintenance* regards non-instrumental activities such as hygiene and maintenance; and (5) *quality of use*, specifically makes space for themes related to the hedonic quality, such as aesthetic experience and stimulation, not only for the person with dementia but for all parties involved; finally, the third part includes (6) *ethics and risk*, which considers not only privacy and safety concerns, but also challenges such as ensuring respect and dignity for the person or concerns related to how products are marketed; (7) *cost and access to the product*; (8) *the inclusion of people with dementia, informal carers and dementia care experts in the design process*; and (9) *sustainability*, related to prolonging the life-time of the product, material choice and product disposal.

The project is now evaluating these lists in the Living Labs. This evaluation will contribute to the development of the final list of considerations, which will eventually result in a certification procedure for products for people living with dementia to receive a mark. Although this is still an ongoing process, thus still no final conclusions can be made, we share a few preliminary points for reflection. After evaluating the selected products within the Living Lab focusing on leisure products, and comparing the reflection of the seven participants, brief specific considerations were added to some of the general themes and sub-themes of the list so far, for instance, regarding the clarity of the product's purpose, information on the possibility and process of personalization, among others. Furthermore, participants also reflected and made suggestions to include different answering options (e.g. scales, checkboxes, answering through observation) in order to make it more accessible during the evaluation procedure. While evaluating the list, and making these additions and suggestions, we realized that several aspects were also having an influence on the evaluation. These aspects were preliminarily organized into two clusters: external (e.g. the caregiver, or requirements of use like Wi-Fi) and internal influences (e.g. the stage of dementia, the person's ability to use and understand of the product's purpose). The impact of these influences on the evaluation will be a topic for further reflection in the course of the project.

The initial Living Lab evaluation of the list can be seen as a pilot project that can be adapted to the evaluation process to receive the *Certification-D* mark. The intention of the development and implementation of this mark is that it will stimulate the production of products for people living with dementia to be able to live longer at home. The resulting final lists will be used twofold: on one hand, they serve as a basis for a reference guide for designers, companies and manufacturers, on the other hand, they can be used to develop scripts for product evaluation.

5 Conclusion and Critical Remark

Initially, the authors of this paper were reluctant in engaging with the *Certification-D* project, namely due to the diversity of products involved, the specific condition of dementia, and to previous failed attempts in creating requirement lists. However, because of the project's ambitious and relevant societal goal, various actions were undertaken to overcome this hesitation and the project's challenges and uncertainties: (a) placing caregivers, people with dementia (and family) and their quality of life central, not only in the initial creation of the lists of considerations, through embedding best practices and values of person-centered dementia care, but also through the setup of Living Labs to involve them in various stages of the project, from the validation and refinement of these lists, to product evaluation, and to the development of the certification procedure; (b) rethinking and reframing design requirements as considerations, including both general and product-specific aspects, and starting from person-centered dementia care and quality of life of people with dementia; (c) a strong engagement of experts from different sectors (care, industry leaders, researchers, designers, people with dementia), through their involvement in the preliminary workshop and on the

Delphi study. At the time of writing, the lists of design considerations are being finalized through their individual evaluations in Living Lab settings, thus giving people with dementia, family and caregivers a voice in the creation of these lists.

Despite the fact that we attempted to overcome these challenges, the actions did not always have the intended result nor did not serve our intended goal. One specific insight concerns the participation of people with dementia. Although we strongly are convinced of a participatory approach, the initial lists that came about after the first preliminary workshop never drastically changed after the Delphi study nor are there any clear indications that the Living Lab evaluation of the list will bring any major alterations. This could mean that the consortium itself (academics, designers, SME's, caregivers, etc.) combined with the expertise of those involved in the Delphi Study was strong enough to come up with a list that is well-defined and described. Interestingly, especially for the leisure list, the first list was detailed and deep, showing a strong body of knowledge, while the market for leisure products for people with dementia is now starting to become well developed. Although it is hard to draw conclusions that go beyond this single case, it might show that the knowledge on design and dementia (be it directly through experts inside of care, design, academia or business or indirectly via literature) is maturing. Consequently, as the first preliminary results of the Living Lab evaluations do not show any significant changes to the structure or the content of the lists, one could question the value of the involvement of people with dementia in this type of study. It is of the utmost importance to stress that the authors strongly believe those affected by design (in this case people with dementia and their family) should be involved in a design's development. However, in this study it is suggested that involvement did not alter or contribute to the results and this may mean that, in some type of studies, involvement should not be facilitated simply as a means to generate better results, but can also come out of purely ideological motivations: it being your duty as a researcher/designer to involve people with dementia in processes which have the potential to impact their current situation or their future.

Acknowledgements The work is supported by Interreg North-West Europe. More information can be found on the project's website: www.nweurope.eu/CERTIFICATION-D. The authors would like to thank all the participants of the Delphi study and Living Lab.

References

1. Kuchler, H.: Amgen, Novartis abandon two alzheimer's drug studies, https://www.ft.com/content/464ecd76-a41f-11e9-974c-ad1c6ab5efd1 (2019)
2. BBC News: Pharma giant ending alzheimer's research, https://www.bbc.com/news/health-42633871 (2018)
3. Livingston, G., Sommerlad, A., Orgeta, V., Costafreda, S.G., Huntley, J., Ames, D., Ballard, C., Banerjee, S., Burns, A., Cohen-Mansfield, J., Cooper, C., Fox, N., Gitlin, L.N., Howard, R., Kales, H.C., Larson, E.B., Ritchie, K., Rockwood, K., Sampson, E.L., Samus, Q., Schneider, L.S., Selbæk, G., Teri, L., Mukadam, N.: Dementia prevention, intervention, and care. The Lancet. **390**, 2673–2734 (2017)
4. Hendriks, N., Slegers, K., Wilkinson, A.: Lessons learned from participatory design in dementia care: placing care partners at the centre. In: Kanstrup, A. M., Bygholm, A., Bertelsen, P., Nøhr, C. (eds.) Participatory Design & Health Information Technology, pp. 63–77 (2017)
5. Hassenzahl, M.: The thing and I (Summer of '17 Remix). In: Funology 2, pp. 17–31. Springer, Cham, Cham (2018)
6. Gowans, G., Dye, R., Alm, N., Vaughan, P., Astell, A.J., Ellis, M.: Designing the interface between dementia patients, caregivers and computer-based intervention. Des. Journal. **10**, 12–23 (2007)
7. Van Rijn, H., van Hoof, J., Stappers, P.J.: Designing leisure products for people with dementia: Developing "'the chitchatters'" game. Am. J. Alzheimer's Dis. Other Dementias. **25**, 74–89 (2009)
8. Anderiesen, H., Scherder, E., Goossens, R., Visch, V., Eggermont, L.: Play experiences for people with alzheimer's disease. Int. J. Design **9**, 155–165 (2015)
9. Alm, N., Astell, A.J., Gowans, G., Dye, R., Ellis, M.P., Vaughan, P., Riley, P.: Engaging multimedia leisure for people with dementia. Gerontechnology **8**, (2009)
10. Robinson, L., Brittain, K., Lindsay, S., Jackson, D., Olivier, P.: Keeping In Touch Everyday (KITE) project: developing assistive technologies with people with dementia and their carers to promote independence. IPG. **21**, 494–502 (2009)
11. Timlin, G., Rysenbry, N.: Design for dementia: improving dining and bedroom environments in care

homes. Helen Hamlyn Centre, Royal College of Art, London (2010)
12. Huldtgren, A., Mertl, F., Vormann, A., Geiger, C.: Reminiscence of people With dementia mediated by multimedia artifacts. Interact. Comput. **29**, 679–696 (2017)
13. Lazar, A., Edasis, C., Piper, A.M.: Supporting people with dementia in digital social sharing. In: CHI '17 Proceedings of the CHI Conference on Human Factors in Computing Systems, pp. 2149–2162. ACM, New York (2017)
14. Mäki, O., Topo, P.: User needs and user requirements of people with dementia: multimedia application for entertainment. In: Topo, P., Östlund, B. (eds.) Dementia, Design and Technology: TIME to Get Involved, pp. 61–75. IOS Press, Amsterdam (2009)
15. Zeisel, J.: Improving person-centered care through effective design (2013)
16. Jakob, A., Collier, L.: How to make a sensory room for people living with dementia: a guide book (2014)
17. Halsall, B., MacDonald, R.: Design for dementia (Volume 1 and 2). The Halsall Lloyd Partnership, Liverpool (2015)
18. Orpwood, R.: Involving people with dementia in the design process: Examples of iterative design. In: Topo, P., Östlund, B. (eds.) Dementia, Design and Technology: Time to Get Involved, pp. 79–95. Dementia, Amsterdam (2009)
19. Hendriks, N., Huybrechts, L., Wilkinson, A., Slegers, K.: Challenges in doing participatory design with people with dementia. In: Winschiers-Theophilus, H., D'Andrea, V., Iversen, O.S. (eds.) PDC '14 Proceedings of the 13th Participatory Design Conference, pp. 33–36. ACM, New York (2014)
20. Lazar, A., Edasis, C., Piper, A.M.: A critical lens on dementia and design in HCI. In: CHI '17 Proceedings of the CHI Conference on Human Factors in Computing Systems. CHI, New York (2017)
21. Morrissey, K., McCarthy, J., Pantidi, N.: The value of experience-centred design approaches in dementia research contexts. In: CHI '17 Proceedings of the CHI Conference on Human Factors in Computing Systems, pp. 1326–1338. ACM, New York (2017)
22. Treadaway, C., Taylor, A., Fennell, J.: Compassionate design for dementia care. Int. J. Des. Creat. Innovation **7**, 144–157 (2018)
23. Branco, R.M., Quental, J., Ribeiro, Ó.: Materializing personhood: design-led perspectives. In: Brankaert, R., Kenning, G. (eds.) HCI and Design in the Context of Dementia, pp. 111–129. Springer, Cham, Cham (2020)
24. Dening, T., Gosling, J., Craven, M., Niedderer, K.: Guidelines for designing with and for people with dementia. MinD—designing with and for people with dementia (2020)
25. Dementia Engagement and Empowerment Project (DEEP): DEEP guides, https://www.dementiavoices.org.uk/deep-guides/
26. Williams, R.: How to design a website for someone affected by dementia, https://www.alzheimers.org.uk/blog/how-design-website-someone-affected-dementia
27. Kennedy, H.P.: Enhancing delphi research: methods and results. J. Adv. Nurs. **45**, 504–511 (2004)
28. Brankaert, R.: Design for dementia: a design-driven living lab approach to involve people with dementia and their context (2016)
29. Ettema, T.P., Dröes, R.-M., de Lange, J., Mellenbergh, G.J., Ribbe, M.W.: A review of quality of life instruments used in dementia. Qual Life Res. **14**, 675–686 (2005)
30. Verloo, H., Salina, A., Fiorentino, A., Cohen, C.: Factors influencing the quality of life perceptions of cognitively impaired older adults in a nursing home and their informal and professional caregivers: a mixed methods study. Clin. Interv. Aging **13**, 2135–2147 (2018)
31. Kitwood, T.: Dementia reconsidered. Open University Press, Maidenhead (1997)
32. Brooker, D.: Person-centred dementia care: making services better. Jessica Kingsley Publishers, London (2007)
33. Hassenzahl, M.: The thing and I: understanding the relationship between user and product. In: Funology 1, pp. 301–313. Springer, Cham, Cham (2003)
34. Hassenzahl, M.: Experience design: technology for all the right reasons. Morgan & Claypool, San Rafael, CA (2010)
35. Gibson, G., Newton, L., Pritchard, G., Finch, T., Brittain, K., Robinson, L.: The provision of assistive technology products and services for people with dementia in the United Kingdom. Dementia **15**, 681–701 (2014)
36. Oxford Learner's Dictionary: Oxford Learner's Dictionary, www.oxfordlearnersdictionaries.com

Designing Conversations: Using Life Stories to Improve Dementia Care

Claire Brophy, Alethea Blackler, Jon Hannan, Caylee Raber, and Vesna Popovic

Abstract

Dementia is a long-term condition that impacts on various aspects of a person's functioning. The person-centred care approach has been shown to be integral to quality of life however the communication challenges facing people with dementia makes understanding the needs of the person difficult. Our research shows that this is a big issue for people with dementia and their families and carers and can be compounded as they navigate different elements of the care. This paper explores the role of design in the context of creating tools to support communication with people living with dementia and personalise their care, with a particular focus on the use of co-design methods for drawing out and sharing people's life experiences and stories. Two case studies are presented which describe how designers and design educators in Australia and Canada have approached this topic.

Keywords

Dementia · Design · Person-centred care

1 Background

Dementia is a term used to describe a group of syndromes, including Alzheimer's Disease, which are characterized by deterioration in cognitive functioning. Dementia affects memory, mood, comprehension and communication and has been identified by the World Health Organization (WHO) [1] as one of the major causes of disability worldwide, impacting on both the person and their caregivers and families.

At present it is estimated that 50 million people have dementia worldwide, and every year a further 7.7 million individuals are diagnosed with the condition [2]. The total estimated worldwide costs of dementia were US$818 billion in 2016 and in 2018 it became a trillion-dollar disease [3] This increase will have an enormous impact on health care and social services globally. For instance, Alzheimer's Australia and the National Centre for Social and Economic Modelling estimated that in 2017 over 400,000 Australians were living with dementia, a cost of $9 billion per annum [4]. By 2056 these numbers are expected to increase to over 1 million individuals, with direct costs rising to over

C. Brophy (✉) · A. Blackler · V. Popovic
Queensland University of Technology, Brisbane, QLD, Australia
e-mail: cj.brophy@qut.edu.au

J. Hannan · C. Raber
Health Design Lab, Emily Carr University of Art + Design, Vancouver, BC, Canada

R. Brankaert et al. (eds.), *Dementia Lab 2021: Supporting Ability Through Design*, Design For Inclusion 2, https://doi.org/10.1007/978-3-030-70293-9_3

$24 billion. In Canada similar figures exist with recent estimates suggesting that the total cost of dementia in Canada is $12 billion CAD per annum [5].

There are over 200 subtypes of dementia and whilst epidemiological studies continue to offer new insights into the disease, to date there is no cure. Emphasis has therefore been placed on the development of approaches and interventions that focus on quality of life and strategies to equip individuals to cope with the challenges that living with this long-term condition brings. However, whilst approaches to caring for people with dementia have changed dramatically over the last few decades, stereotypes and false beliefs are still too prevalent [6]. Lack of awareness and understanding often leads to stigmatisation, reduced quality of life and barriers to quality care. Unfortunately, many people, professionals included, have difficulty seeing the person (and not just the patient), behind the symptoms.

Many of these stereotypes and pre-conceptions need to be challenged. Despite the cognitive decline in dementia, there is evidence to suggest that aspects of self-identity are not lost in the person, that a person's identify is intact and perhaps just concealed [7–9]. Person-centred care is an approach to care that acknowledges the importance of self and identity—regardless of cognitive ability [9]. In his pioneering work on the theoretical perspective of person-centred care, Kitwood defined this notion of identity as personhood; "a status or standing bestowed upon one human being, by others, in the context of social relationship and social being. It implies recognition, respect and trust" [10, p. 8]. Some attempts have also been made to incorporate person-centred care into patient files with documents such as Dementia UK's Life Story Template [11] and the This is Me leaflet from the Royal College of Nursing and the Alzheimer's Society [12]. These documents are "intended to provide professionals with information about the person with dementia as an individual" [12, p. 4], thereby enhancing care and support in an unfamiliar environment. Unfortunately, the research reported here found that these types of documents are not effective and/or not used effectively.

Kitwood's work has made a significant impact on dementia care [13], particularly around the introduction of non-pharmacological tools used in care environments. Based on various perspectives and methodologies, non-pharmacological approaches address loneliness and boredom, encourage social interaction and increase quality of life [14]. These approaches focus on the needs of the person and recognise that behaviours are influenced by social and emotional needs as well as past experiences and life histories. Assisted-living caregivers using person-centred knowledge such as life histories were able to anticipate the needs of individuals and customize their care [15]. The ability of staff to provide higher quality care improved the quality of life for people in care, and in turn, increased staff job satisfaction. While there is no consensus as to which interventions are most effective, creative approaches such as art therapy, music, and life story work have developed significantly [16]. Research continually points to the potential of such psychosocial interventions at improving self-esteem, enhancing cognition and physical capability, and reducing behavioural problems [17–20].

Person-centred care has featured heavily in healthcare literature for some time. However, there is a growing recognition of the challenges to the implementation of such an approach. In part this is because the medically modelled healthcare industry has largely ignored the role of personal histories and social contexts in dementia care [21]. Caregivers noted that, rather than being included as a part of their training, person-centred knowledge came from experience, where information was passed from long-term staff to new caregivers [15]. A British study by Popham and Orrell [22] found that in residential facilities, person-centred care was seen as an "ideal rather than realistic goal" [p. 185].

While there is evidence to suggest an increased recognition of person-centred approaches across the broader healthcare industry [23] actual day-to-day implementation remains challenging and inconsistent [24]. Some tools, such as Dementia Care Mapping (DCM) [25], have been criticised as being expensive, time consuming and somewhat out of reach for the basic grassroots level of care [26, 27]. Significantly, in their critical comparative review of published tools measuring person-centred care, Edvardsson and Innes [27] concluded that the voice of the person with dementia is still largely absent in the available tools, and that further work is required on tools designed to incorporate the knowledge of people with dementia.

This to some degree is not surprising. This lack of understanding is not without good reason as there are significant methodological challenges in relation to capturing the voices of people living with dementia. Communication has been cited as possibly the most difficult obstacle to overcome when trying to capture the voices of people living with dementia [28]. Dementia impacts communication. Dysphasia is a condition that occurs as a consequence of the neurological changes occurring in the brain and this impacts comprehension and expression of the written and spoken word [29]. Dysarthria, again resulting from neurological damage, can also affect the mechanisms and subsequently the quality of speech. For individuals with cognitive impairment and dementia, additional challenges can include word finding problems, understanding abstract concepts and memory recall [30]. These challenges can be compounded by general age-related changes in sensory acuity. For instance, one study by [31] found that half of people in nursing homes did not communicate with others in the setting because of hearing and speech problems.

Communication problems can pose a significant barrier to understanding the views of people with dementia, although it is very important to continue trying as it is possible to elicit their views [32]. This becomes even more pressing in the context of people living with dementia since individuals can find themselves in numerous care environments. Poor co-operation and communication between different levels of care and from different services affects the quality of care for people with dementia.

2 Transitions

The decision to place a loved one in full time care is perhaps one of the most difficult phases of dementia and can cause overwhelming stress for caregivers. Carers report experiencing conflicting emotions such as sadness and resentment alongside relief and comfort [33]. There is a common sense of becoming homeless during the transition to long-term care. Even for people without significant cognitive impairment this change requires extensive negotiation of the new surroundings; they need to fit in, learn the rules and create a place for themselves [34]. For new people living in long-term care with dementia there is often a feeling of abandonment and a sense that perhaps they have done something wrong; they cannot understand why their loved ones come to visit but refuse to take them home [10].

Heliker and Scholler-Jaquish's study [34] demonstrated that sharing of personal stories and lived experiences helped people in care illustrate what is meaningful and important to them. People used stories to provide context for their lives, "to preserve their history and culture, to validate their self-identity, and to form bonds with others" [34, p. 41]. As a result of their study, story sharing is a strategy being developed and encouraged during the transition process. From admission, transitional care teams including the people living in care, their family members, healthcare professionals, other appropriate personnel, participate in story sharing activities with the intention of getting to know the individual and helping them integrate into the new facility.

Reuss et al. [33] presented a model of seven individual factors that contribute to, and are necessary for, positive transitions to full time care. The authors emphasise the need for transparency in decision making to avoid feelings of the process being 'expert-driven'. Their detailed model builds on previous work exploring the process of navigating transitions in care. It demonstrates the value and importance of including the family in the overall process, and further highlights the critical role of strong, open communication between staff, families and advocates for "pre-placement education programs" [33, p. 38]. However, nowhere in this extensive and thorough model is the inclusion of the individuals' life history or personal stories.

3 A Design Approach

Although professional and personal caregivers recognize that having person-centred knowledge such as life histories increased their ability to provide quality care, the use of such knowledge is seen as ideal rather than standard practice. Current methods of collecting and using this information are inconsistent and ineffective—challenges only compounded by the communication difficulties experienced by people living with dementia. With this in mind, our research employed design approaches and methods aimed at improving the collection and use of person-centred knowledge in order to improve communication and personalization in dementia care.

Co-design methods are intended to draw out latent and tacit knowledge through creative, hands-on activities, often referred to as "make tools". By using creative methods of engagement and actively involving people, or users, in the design process, these tools and methods are similarly useful at drawing out people's stories and experiences. Co-design activities typically take the form of word/image association, quizzes, games, or hands on design or making in the moment [35], such as painting, drawing and construction activities, all with an added storytelling element. Such methods elicit deeper forms of expression and encourage people to share their experiences, thoughts and ideas [36].

This paper now describes two case studies, one from Australia, one from Canada, which developed approaches to sharing and recording the stories of those entering and living in aged care facilities. Both studies demonstrate the value of design methods; both emphasise the value of personal life histories, storytelling, and *knowing the person*, to the delivery of quality in dementia care.

4 Case Study 1—Knowing the Person

In Case Study 1 we present research that sought to understand the application of person-centred care, and person-centred knowledge in particular, in the various environments that a person with dementia would experience across the care system. The research illustrates the types of knowledge used both by professional and family caregivers; how and where this knowledge is used, and where it is not; as well as opportunities to improve the use of person-centred knowledge in dementia care.

4.1 Method

Case study 1 was based in South-East Queensland, Australia, however, data were collected from both local and international sources. The research, approved by the QUT Human Research Ethics Committee as part of the curriculum, was conducted in three phases—(1) expert interviews; (2) online content analysis, and (3) unobtrusive observations.

In Stage 1, four interviews were conducted with health-care professionals. Participants were approached from within the researchers personal network and selected to represent various environments and experiences of the care spectrum and service delivery: two experts in dementia care (email interview with a Director of Assisted Living and Memory Care at a continuing care facility in the United States; a Manager from Alzheimer's Australia's Dementia Behaviour Management Advisory Service) and two front-line staff in the health care industry: a Nurse Unit Manager (NUM) of an emergency ward at a Brisbane hospital and a paramedic with the Queensland Ambulance Service. Audio recordings of the three face-to-face interviews were transcribed verbatim; in the case of the email interview, the written correspondence was used in place of a transcript.

In Stage 2, content analysis was carried out on an online community discussion forum supported by the Alzheimer's Society [37]. The forum represents both people with dementia and their caregivers. At the time of the review the site had over twenty thousand members. As of 2020 it had grown to over sixty thousand members. The researcher introduced herself and the project to the site's administrators before any analysis was carried out. This method sought to understand the experiences of carers and people living with dementia first-hand. Our intention though, was on generating a broad, higher-level understanding and we did not record any personal details of the forum's members and do not report on members' individual situations.

In Stage 3, unobtrusive observations were carried out on five separate occasions at a nursing home facility in South-East Queensland in Australia as part of a pre-existing volunteer program. The home, an older facility that was later closed, had 120 residents in four interconnected units: two residential units, one secure dementia unit and one secure mental health unit. Results from each stage are presented below.

4.2 Results

Stage 1: Interviews. Interviews focused on the characteristics and challenges within dementia care. As described by Saldana [38, 39], analysis of interviews was undertaken by following an iterative and inductive coding process to identify recurring themes of consistent meaning and context. Table 1 outlines the categorization of findings into six topics and includes sub-themes to clarify the definition of each. All six topics are briefly discussed here, including quotes from participants.

Topic 1. Needs recognition; this includes emotional, environmental and daily activity needs of people with dementia, the staff and the caregivers. Both experts suggested that understanding individual resident needs proved beneficial in personalizing care and that past life experience had a major influence on how the person coped with the disease.

[Director]: "Everyone with dementia is not the same - 70–80 years of past life plays a huge role in how the individual handles the disease."

Topic 2. The importance of knowledge & knowledge transfer in care environments includes the knowledge staff have about dementia in general and about the person with dementia specifically. Echoing the results presented by [15], participants noted that personal patient information was useful in accomplishing daily work activities and contributed to increased job satisfaction.

[NUM]: "[what's] really good is when you get the photocopy from the nursing home of the drugs that they're on... they write at the top, 'Betty likes her tablets to be crushed, and given with green jelly'... some of them write that, and if I didn't know that I would be trying to shove a massive tablet into Betty's mouth and she would be spitting it in my face".

While no specific reference was made to person-centred care by the researcher during interviews, all participants referred to its use, and to the value of such information when dealing with things such as behavioural issues.

[Paramedic]: "Some dementia patients just want to talk about fish... and next thing you know you have a person that's fighting and screaming and saying they don't want to go anywhere and then you say 'oh, you should have seen this fish that I caught the other day' and they'll stop... 'was it this?, was it this?' and then they'll follow you...".

Knowledge transfer in dementia care occurs at many levels and it can be represented by various types of knowledge. For instance, it may be about sharing knowledge about dementia at a broader societal level in order to reduce the stigma, or it may be detailed personal information about one person with dementia. This knowledge transfer is present from research to industry, from families to facilities, between facilities, and between individual carers and staff within facilitates.

The Manager cited transitions to permanent care as one of the most difficult times for the person with dementia and their family. She stressed that behavioural issues occurring as a

Table 1 Key themes from interview data

	Topic	Sub-topic
1	Need recognition (person, staff, and caregiver)	Emotional: security, familiarity, freedom, preferences Environmental: familiarity, visual cues/triggers, personalisation, engagement Activities of Daily Living: bathing, feeding, dressing, mobility, communication, medications Family/Caregiver need/Staff need/Professionals Emergency situations (immediate needs)
2	Importance of knowledge & knowledge transfer	*Knowledge* Staff knowledge about disease and person Staff attitudes: task oriented vs. person-centred Collecting personal information Availability, storage & accessibility, documentation, formatting, presentation, admission *Knowledge Transfer* Handovers: between facilities, staff to staff, Family to facility, Family to Emergency staff Sharing, teaching, educating Research to industry Cultural/attitude shift—stigmas
3	Influence and suitability of environment	Physical & social environment Treatments, Programs & Activities Policy & Legislation—Health & safety
4	Impact of dementia on caring	Different stages becoming difficult to manage, indicating need for higher care accommodation Reverting back to instincts Different stages as an indication of physical ability, communication and care needs Impact & Identification of Behaviours (physical/emotional) Physical: wandering, aggression, violence, spitting, biting Emotional: agitation, depression, boredom, anxiety
5	Recurring problems restricting quality care	Restriction of Finances & Time: Cost, Affordability, "Big business" Not enough time Staff change over, Temporary staff Lack of necessary information, Lack of knowledge/training
6	Provision of Medical Information	Documentation, medical history, medication, allergies, abilities, cognitive state, mobility, DOB, next of kin, drug

result of this change could be better managed if appropriate information was gathered beforehand.

[Manager]: "It's really the relationship between the facility and the family. That relationship needs to be so healthy for the whole journey within residential care, and it starts before admission."

In exploring knowledge transfer during transitions, participants had high opinions of nursing homes that were able to provide extensive patient and, in the case of emergency response, incident information.

[NUM]: "…and really good nursing homes, they bring a copy of: 'Gertrude likes classical music, she used to have a dog, she needs everything doing when she has her wash, but she likes to brush her own hair', like the really good ones send all of that… the sh*t ones you get nothing, not even an allergy…".

However, while the sharing of detailed patient information is highly regarded and useful, participants of this study noted the size of the files and disorganized nature of documentation was difficult to manage effectively.

[Paramedic]: "…sometime its [personal patient information] documented if you can flick through the 25 pages of scribbled hand writing, photocopied-multiple-times-paper that's handed to you."

Topic 3. The influence and suitability of environments reflects both physical and social environments, from legislation and design, to programs and activities. Interestingly, the results reflect the nature and role of each participant. For example, the paramedic made mention of the impact physical features of facilities had on being able to access and extract the patient in emergency response situation. The director and the manager, both with extensive experience in residential facilities, had holistic perspectives on environments, seeing them as vital to overall quality of care.

[Director]: "Environment plays a HUGE role in the care of someone with dementia…So what has to happen is provide the right environment where they feel like they are home and there are activities that keep them engaged, this redirects them from their fears. The best environments are those that allow freedom to wander within safe limits."

Topic 4. The impact of dementia on caring addresses the identification and impact of both physical and emotional behaviours expressed by the person with dementia; from wandering and spitting, to depression and anxiety. The Director explained that there was no standard for navigating different stages of care and that changes often happened in crisis situations.

[Director]: "The spouse who is the caregiver often becomes ill and is hospitalized, therefore the person with dementia ends up in a care facility in an emergency situation, unless children take over the care. Or, the spouse dies, resulting in admission. Often the spouse caregiver has been covering up and children don't realize how far the dementia has progressed, until they have to take care of mum or dad. If there is no spouse and children are the responsible parties, they tend to be more willing to place mum or dad in a facility."

Topic 5. Recurring problems addresses barriers to providing quality care, fundamental challenges such as financial issues, the affordability of care, and the unfortunate notion of 'get what you pay for'. Participants of this study indicated that the use of temporary and agency staff in nursing homes further impaired the provision of quality care as temporary staff have minimal knowledge about their residents,

[NUM]: "…you have other nursing homes where they just can't keep staff, so the staff are brand new every day, or they're agency every day, and they just don't know the patients, so they just don't know anything about them…".

Other barriers included time constraints and high workloads in care environments, a lack of information, or poor management of information, and a general lack of dementia-specific training across the wider healthcare industry.

Researcher to Paramedic: Do you get any dementia training?

[Paramedic]: "We get a vulnerable clients training, which is pretty piss weak really, it's a chalk and talk… they stand up in front of you for two hours…. blah blah blah… there's no going out to an actual dementia wing and spending time there…".

In line with the literature, staff expressed feelings of frustration and inadequacy at not being able to deliver quality, personalized care.

[NUM]: "… most of them[nurses], go into it because they want to help people, and if they could spend an hour with each one they'd love that because, they'd go home and feel like they'd done their job properly and that's why they went into nursing and I think a lot of nurses go home at the end of the day and feel like they didn't do what they wanted to do…".

Topic 6. The provision of medical information, fundamental documentation such as DOB, allergies, next of kin, etc. Information that is sometimes provided, but often not.

[Paramedic]: "there should be a list for every patient, in their room on their wall…here's their medical history, date of birth, allergies, the medications they're on, that's it… that would then open up a whole different scope of practice for what I can do for that patient…".

The analysis of interview data highlighted the issues of communication that people with dementia face, particularly in relation to their past lives and sense of self. Analysis also pointed to the valuable role that this knowledge of past life may play in communication and personalizing care. The following section presents results from Stage 2, thematic analysis carried out on an online community discussion forum.

Stage 2: Online Forum. The online investigation was conducted in the most popular discussion thread on the website of the Alzheimer's Society [37], which at the time had 23,000 threads and over 280,000 posts compared with an average of 530 threads in other fora within this section. Each page displayed 25 threads and six consecutive pages were reviewed on three separate occasions, totaling 450 threads. To simplify the process, only the initial post of each thread was reviewed to identify key words that would help determine the general topic and the intention of the post.

Most posts were related to sharing experiences and seeking the advice of people in similar situations, including the process of diagnosis and dealing with behavioural and personality changes. In a process similar to the analysis of interview transcripts in stage 1, and with consideration of the themes identified in that data (Table 1), recurring topics were grouped according to similar meaning and context. Eight key topics were identified: (1) Facilities; (2) Diagnosis; (3) Seeking Support; (4) Caring at home; (5) Behaviours; (6) General Sharing; (7) Medication; and (8) Venting.

Analysis illustrated a consistent pattern across the 450 threads reviewed (Fig. 1), of those, 25% (112/450 threads) related to experiences with Facilities and transitions through various stages of care. This topic often generated a substantial response from other members of the forum, was the most prominent in all three separate reviews and will be the focus of the remainder of the research presented here.

Posts about experiences with facilities related to the search for suitable facilities, transitioning a loved one into permanent care and the difficulties experienced by carers during this time. Other notable topics included: a fear of hospitals; a general lack of knowledge, guidance and structure within systems of care; and a recognition that the use of personal information had a positive impact on the care of the person living with dementia.

Family members were often the primary source of care. Care within the home was generally relinquished as a result of the situation becoming unmanageable, where injury had occurred or was a serious risk. Carers seemed unprepared for this point in care and were often asking for "some advice asap". They were confused by the process, which often involved a stay in a hospital or assessment centre while determining what type of facility would be most suitable.

Reflecting the literature [33], the process of transition to permanent care was recognized as one of the hardest parts of the dementia journey. People described being "a nervous wreck" and "panicked" waiting for the call from the hospital to say that their loved one had left for the care home, often realizing that they "had no choice". Forum members expressed an overwhelming sense of responsibility and a mixed sense of guilt and relief when placing their loved one in permanent care. Some felt "awful, sick". One went so far as to say, "at this moment I wish I were dead". Interestingly, within the search for suitable care homes, design features were not discussed as regularly as the attitudes of staff and their interactions with the people living there.

People appreciated staff that were kind, patient, and placed great emphasis on meeting the needs of the individual. Whether in hospitals or nursing homes, carers were aware that personal knowledge about their loved ones had a positive impact on their care and had high opinions of facilities that recognized individual needs. Carers expressed a desire to be consulted about their loved one's care and were keen to provide information that may assist in the personalization of their care. However, they were frustrated that even when personal information was provided, it was not given proper attention, lost or "just thrown away".

Trips to hospital were frequently portrayed as negative experiences, where issues of time and workload within such structured care

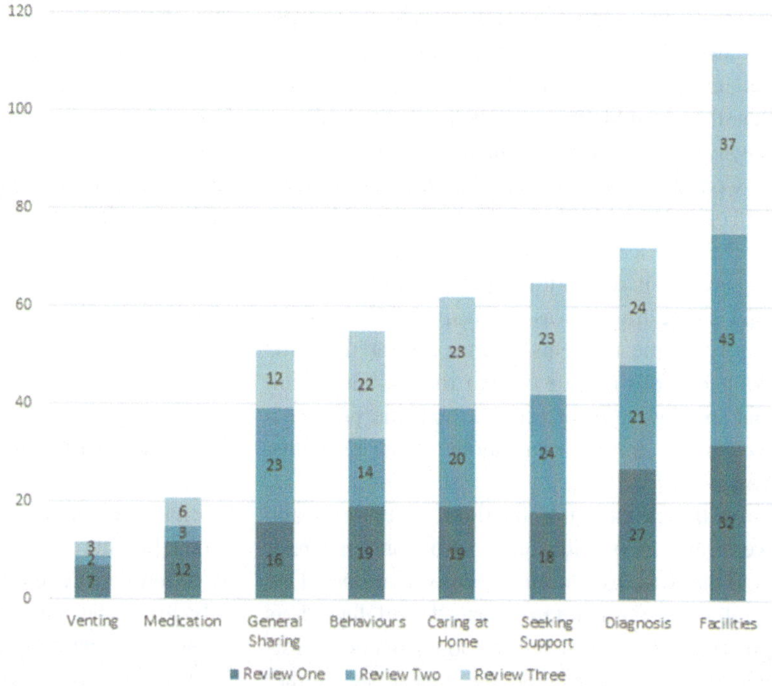

Fig. 1 Frequency of themes in online data

environments appeared to limit interactions and the use of personal information. Many people wanted hospital staff to find out about the person with dementia, not just medical needs but social needs, habits etc. A few minutes to uncover this information from family or carers on admission could make a lot of difference. Comparable with results in Stage 1 (Topic 5) members expressed frustration that staff in both residential and acute care environments did not have even basic training in Dementia care.

Thematic analysis of the online data suggests the journey through care with someone with dementia is worsened by a general lack of knowledge, guidance and structure within systems of care. Similar to the information provided by the Director in Study 1, transfers to permanent, long-term care often came at the last minute, as a result of rapid deterioration, a crisis, or where care at home was no longer manageable.

Given the overwhelming popularity such of forums it seems the information and support carers are seeking is not readily available to them elsewhere. Seeking information and advice from such an avenue also suggests that caregivers are more likely to trust the opinions of those that have had firsthand experience in caring for a loved one with dementia. The next section presents analysis from the final study of this project, unobtrusive observations at a residential nursing home.

Stage 3: Unobtrusive Observations. As part of a pre-existing volunteer program, unobtrusive observations were carried out on five separate occasions at one residential nursing home. The visits occurred on Sunday mornings between 9 and 11 am, and were the only activity scheduled over the weekend. This timeslot may have reflected a general lack of activity within the facility. It should be noted that this was an older facility and the unit visited most was providing higher-level residential care. Residents had very limited independent mobility and most were experiencing mild to moderate cognitive decline. There was some interaction with residents in the communal activity/dining room, however, most residents were visited individually in their rooms and all, apart from one, were in bed. During an orientation to the facility, the facility's volunteer and activity coordinators identified which residents to approach and which were best avoided. The coordinators discussed a plan to collate some basic information about residents that they would then supply to the volunteers; however, this did not eventuate.

Contrary to advice from the Director in Study 1 (Topic 3) and [40], that facilities should strive to feel like a home, most rooms visited by the researcher had minimal personalization, with staff suggesting it was the responsibility of the family. Some residents' names appeared on the doors or walls outside of their rooms, but not all, and some were incorrect. The staff did possess a wealth of information about their residents, however.

Staff facilitated introductions to residents and were able to provide information on each resident that was used to initiate conversations, "Eric used to live on a farm" or, "Hazel had dogs growing up". When this personal history information was gained by the researcher, either through conversations with staff or by artefacts in residents' rooms, it was effective in helping stimulate conversations with residents. In several instances, residents would become much more lucid, animated and conversational when asked personal questions.

Where a resident's speech was difficult or limited, staff were able to 'translate' some of the conversation during these visits. Staff were also able to recognize when a resident was uncomfortable and shift their position, even though the resident was not verbally able to communicate this discomfort.

Unfortunately, however, without staff on hand, life history information of residents was not provided or readily available. Staff suggested this lack of information was often due to the acute nature of transitions in dementia care. Ultimately, most families were not prepared. Staff expressed disappointment, even sadness, at the fact that some residents did not have any family to help create this personal space.

Over the course of five visits it became apparent that the physical appearance of the old building played far less a role in the care of the residents when compared to the knowledge used by the staff during interactions with residents. The staff were aware of residents' preferences, interests and backgrounds; they showed sensitivity to their needs, desires and physical comfort. They encouraged interaction but also recognized when it was not appropriate. It was clear staff had the welfare of their residents as a priority. Unfortunately, however, successful volunteer interactions would not have been possible without the staff escort. This suggests that current methods of collecting and sharing this vital information are not effective.

4.3 Summary of Findings

Overall, the findings of Study 1 agree with the extant literature. Issues surrounding experiences in different care environments featured repeatedly in all the data. Changes from independent living to a permanent care facility, or from a facility into an acute care environment, are challenging and painful experiences for both the person with dementia and their carers.

There was overlap in the identified themes across the data and although it was difficult to distinguish areas of priority, the findings were developed into three overarching themes: (1) Communication and Engagement; (2) Recognition and Awareness; and (3) Knowledge and Compassion. Communication and Engagement emphasizes the need for improved communication between care providers, and increased engagement with the person and their family during transitions to and within care. Recognition and Awareness is about recognizing the impact that an individual's life history has on their needs and behaviours. Knowledge and Compassion is the need for increased understanding about dementia in general across all care providers.

Limitations are recognized regarding the data presented. Given that the online analysis was conducted on consecutive pages, one particular thread may have influenced the appearance of similar discussion topics and future studies might consider a review of more than one forum. Observations were only conducted on Sunday mornings within one high-care nursing facility where residents had limited communication and mobility. It is suggested observations within a facility where residents have more independence and awareness may have provided a more balanced view. Lastly, data for Study 1 has been collected from three continents. It is recognized that issues and processes of one care system may not apply to that of another. However, the general context of poor communication, a lack of training and knowledge, and attention to individual needs appeared in all data regardless of origin.

The next section describes a product concept based on the outcomes of the research in case study 1. With the aim of improving the availability of personal information about people living with dementia for care providers and services, and promote a person-centred approach to dementia care, the concept aims to create a more coordinated system of communication, focused on the individual.

4.4 Person-Centred Design Development of a Reminiscence and Conversation Kit

In considering design solutions the concept presented here specifically sought to improve the process of collecting the life history information in the first place. Given the progressive nature of dementia, a system focused on information collection would need to be provided to the person with dementia in the early stages of the disease, while communication is still possible and they are able to consent to sharing their stories.

ReCollect (Fig. 2) is a personal reminiscence and conversation kit that is intended to be given to the person on diagnosis. Designed to reflect core family leisure activities and stimulate conversation, the concept provides for the collection, storage and display of small objects of significance. Most importantly, a set of cards provides ten themes with which to begin exploring a person's life history (Fig. 3).

Each card has a question on the front and space for answers on the back. The aim of the cards is to stimulate conversation and quality interaction between the person, their family and close friends in the early stages of the disease, and before significant cognitive decline.

In later stages and after permanent placement in care, the cards provide valuable information that can be easily and quickly accessed by carers and nursing home staff (Fig. 4). This information provides starting points for conversation and learning over time.

The questions and themes of each of the cards (Table 2) were derived from several sources; the essential information included in documents such

Fig. 2 The recollect concept

Fig. 3 Examples of the cards

as Dementia UK's Life Story Template [11]; the Alzheimer's Society's This is Me leaflet [12]; and our earlier findings outlined in Table 1 such as emotional needs and familiarity, environmental preferences, as well as daily activities and routines. Importantly, as a low-cost and accessible concept, the cards also address some of the challenges identified in Table 1, particularly around financial restrictions and knowledge sharing across the care system.

To elicit some of those needs, the card themes and questions have been guided by autobiographical interviews [41–44]. Autobiographical memory is related to both episodic memory—information at a specific time and space, and semantic knowledge of the self—like knowing where and when you were born [43]. Finally, inspiration was taken from examples in published articles by Svoboda [43] and Concordia University [45], as well as the Smithsonian Institution's Folklife and Oral History Interviewing Guide [46].

The purpose of ReCollect is to encourage families to document the life of the person with dementia in order to improve the quality of their long-term care. Feedback from initial trials of the cards has been positive. One family offered the following comments, "…thought you might like to know that we have all been using the cards for Granny and they are proving quite distracting! Once she starts to think about an answer, it leads her onto something else and thus another topic of conversation. Can easily pass half to three quarters of an hour! Some of the questions are quite emotive so we have to be a bit selective at times! - depending on her mood… I think they are an excellent tool for those with Alzheimer's or dementia" (A. Baron. personal communication. June 2012). Future work within this concept should revise and refine the topics and questions included in the first iteration, especially a more in-depth understanding of the specifics of health care documentation.

Building on, and applying, the importance of life histories and of knowing the resident demonstrated by Case Study 1, Case Study 2 presents a codesign method as a way of engaging people living in care homes in the creation of artefacts to communicate their life experiences.

5 Case Study 2—Perspectives

Perspectives is an intergenerational teaching program that brings together art and design students with people in long-term care to co-design and co-create outcomes that reflect both the participants' life experience and the shared

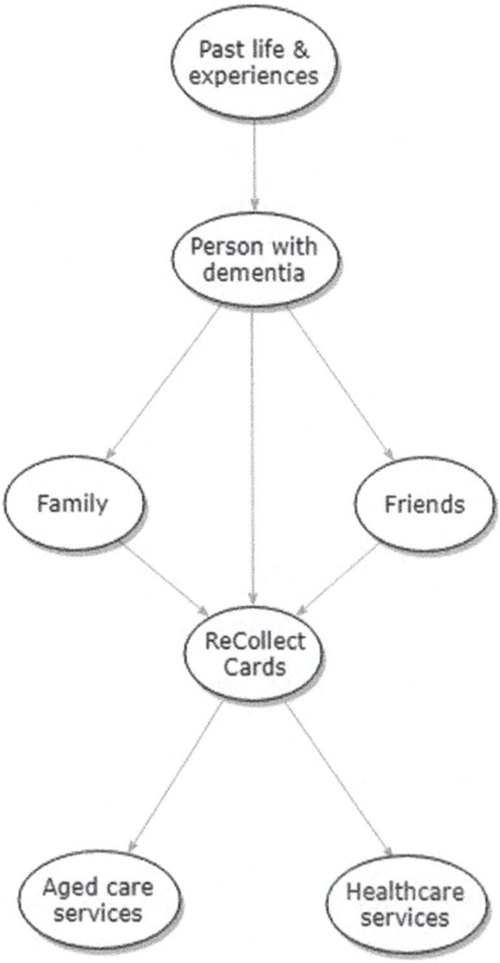

Fig. 4 Card interactions

experience of the intergenerational exchange between two groups of people at different stages in their lives.

Developed at Emily Carr University, Vancouver, Perspectives, formerly called Zeitgeist, takes place over the course of twelve weeks and sees students and people living in care form small groups to undertake a series of co-creational activities, designed by the students, that emphasise the importance of storytelling and sharing [47]. The materials generated during these sessions form the basis of the program's outcomes, with these outcomes typically taking the form of high quality, printed publications that are then shared between participants, their families and care home staff at a storytelling and sharing session at the end of the program.

5.1 Relationship Building & Identity

In many local care homes a 'Getting to Know Me' form (similar to This is Me leaflet from the Alzheimer's Society) is the main tool used to capture a person's history and character upon entering a long-term care home plays a large role in determining the care they will receive. The form is poorly designed from both a visual design and user experience perspective, which could easily leave someone feeling like their life experience has been edited down to a few fields of text. Little care is given to opportunities for personalisation and it is not possible to differentiate the form to accommodate different life experiences. Many people entering long-term care are doing so because they have been diagnosed with dementia, and it is not uncommon for these people to already have low self-esteem and to feel like they have a reduced role in society following their diagnosis [48]. Further, this form is often used upon transition into a home and is therefore collected and used at a single point in time.

To prepare communication design students to lead co-design activities with participants living in long-term care (a context which is usually unfamiliar to students), we begin with a small introductory assignment which asks students to re-design the 'Getting to Know Me' form.

The students are first asked to put themselves in the position of someone entering a care home and complete the form with one of their peers. This quickly leads to frustration, boredom due to the repetitive nature of the form and an overwhelming sense of empathy from the student as to the challenge someone faces in this situation. The act of completing the form often triggers conversations around what it takes to truly get to know somebody, touching on themes of trust, respect and divulging personal information to people who we do not know. The students are challenged to reimagine the experience of getting

Table 2 Card themes, topics and example questions

	Card theme	Topics & questions
1	Growing up	School/teachers Childhood food/toys/books/games/pets Complete the following sentence: The words that would best describe me as a child are…
2	Becoming an adult	How old were you when you left home? Describe the different types of jobs you have had over the years… What advice would you share about marriage?
3	Family	Do you have any favourite family occasions /traditions? How would you describe your parents? Are there any family heirlooms, recipes, or other things that have been passed down in you family?
4	Sports	Do you have a single greatest sporting event of all time? Do you have a favourite sport to watch? Did you play any sport
5	Entertainment & events	Moments in life/traveling… Describe your perfect party Are there any significant world events that you remember?
6	Faith	Values & beliefs How have your beliefs changed or remained the same over the years? Describe your faith.
7	Love	Do you remember your first kiss? What does love mean to you?
8	Your Life Now	Routines/habits/activities Do you have a morning routine? Can you walk us through it? Do you collect anything? Complete the following sentence: On the weekends I like to…
9	The Little Things	Personality/idiosyncrasies Is there anything you are afraid of? What is your life are you grateful for? Is there something that brings you comfort?
10	If…	Imagination/fantasy If you could have dinner with anyone, who would it be? If you could have stayed one age forever, what would it be? If you could have been at any event in history, what one would you go to?

to know someone and propose alternative solutions to the form and the experience of a care worker getting to know someone. They typically start by restructuring the form from a visual perspective, but this leads to broader conversations of what may be more appropriate ways of approaching this task.

The completion of this assignment coincides with the first meeting of students and care home participants. In the first meeting groups of 2–3 students and 2–3 care home participants are formed—groups which they will remain in over the course of 6 visits, in order to gather, document and share stories by ultimately co-designing a publication. In the first visit, the process of getting to know each other begins with a small icebreaker led by the care home recreation therapist (Fig. 5).

5.2 Co-creation & Personalization

Having met the people in their groups once, and started rapport building, students are asked to

Fig. 5 Students and care home participants (and Elvis) engage in ice-breaker activities

design a series of activities that will engage them in a process of story-telling and story-sharing with the ultimate goal of creating a publication together that features these stories. Students must creatively design participatory activities that are personalized to the interests and experience of the people living in care, while accommodating individual abilities, allowing for maximum participation where possible. It soon becomes apparent that although many of the participants may have lost a degree of cognitive ability or have physical limitations, their personality and personhood remain intact [49]. There is an emphasis on designing activities that are focused on an individual's abilities rather than disabilities, agreeing on a shared language and discussing goals and objectives together to, where possible, create a valuable experience for all participants [50].

This personalized focus is often the key to the unique aspect of the group outcomes, as individual character traits and interests become the focal point of the exploration. Through the process of making, participants can express preferences, feelings and thoughts that would have been difficult for them to verbally express, with co-design methods drawing on more tacit knowledge than traditional approaches. These personalized 'things' allow for them to construct their own unique creative language with which to express themselves, enabling people living in care to act as designers, while also helping to level the power imbalance that can exist when 'jargon' is used by the students. This staged and personalized co-design approach also allows participants to see their contribution as the project develops and trace their input back to the beginning [51].

For example, one group of students created a large set of dice to play with to accommodate the visual impairment of one of their group members. Rolling the die then corresponded with a question that was to be answered by the roller. In this game both the people living in care and the students participated in sharing stories about themselves by rolling and answering questions; it was not a one-way interview. The simple act of creating larger dice for the game was met with great enthusiasm by one participant who excitedly exclaimed that she could see the numbers! This customization enabled the individual to participate fully by recognizing her abilities.

In another group, several of the members had expressed an interest in bowling as a pastime in the home. This led students to develop an activity that took the form of a bowling competition, with different storytelling questions associated with different scores (Fig. 6).

Students are instructed to "scaffold" activities to aid participation and discuss how the materials

Fig. 6 An interactive bowling game is used as a game to get to know each other

generated will be used following the sessions [52]. This ongoing collaboration is an important factor, as students are asked to leave activities "open" and prototypical in nature, allowing for quick customization following feedback and therefore personalization by individuals in the home [53] (Fig. 7). The participatory nature of this assignment allows people living in care homes to become actors in the design process and not just passive observers [54] (Fig. 8) Through these storytelling activities participants find common ground, often through shared histories of places, cultural backgrounds and preferences for art, music or sports [55].

In this context student interactions with people living in care differ from typical staff interactions, engaging creatively with the explicit goal of drawing out their story, and as a result students are often are able to uncover new stories and information about the people living there that staff had not previously known, even from individuals who had been living in homes for long periods of time already.

Fig. 7 A collage activity is used to support storytelling

Fig. 8 Students and people in care review publication content

5.3 Storytelling & Validation

Materials generated through the co-design activity sessions are then collated and refined by students into publications that can be returned to participants and shared with their families and care home staff (Figs. 9, 10, and 11). This typically sees the groups telling stories together and participants expressing a sense of pride in what they have created. There is often a performative aspect of the storytelling session, with live readings from people living in care and even instances of individuals singing songs and performing to all groups. Dementia often means a person goes through "biographical destruction", seeing their identity and image restricted [56]. This celebration of individual stories allows participants to celebrate their experience and identity, providing a great sense of pride, and also allows them to actively contribute to solving their own problems surrounding identity and self-expression [57].

Having run the project four times, there have been opportunities for some people living in care to repeat the program multiple times. One participant even returned for a third time and proudly displayed her two previous program outcomes in the first session with her new group. In this instance she became the expert designer, using her previous experience to inform students on how to best proceed with their co-creational activities [58]. This also demonstrates that getting to know someone takes time and is never complete. Filling out a 'Getting to Know Me' form at the point of transitioning into a care home will not sufficiently enable to staff to know a person. Opportunities for ongoing story-sharing can be valuable.

5.4 The Research Study

A qualitative research study was conducted with ethics approval by Emily Carr University (Research ethics #: 100246) and the University of British Columbia (UBC; research ethics #: H18-01781), seeking to explore the impact of the Perspectives program for the participants.

5.5 Method

The study consisted of interviews with long-term care participants, a pre- and post-survey with students, and an observational study during the program. The observational study involved making detailed notes regarding engagement and interactions within the sessions. Care home staff were also involved in mid and post program interviews, during which they were asked about

Fig. 9 Example of a group's finished publication in the form of a comic book

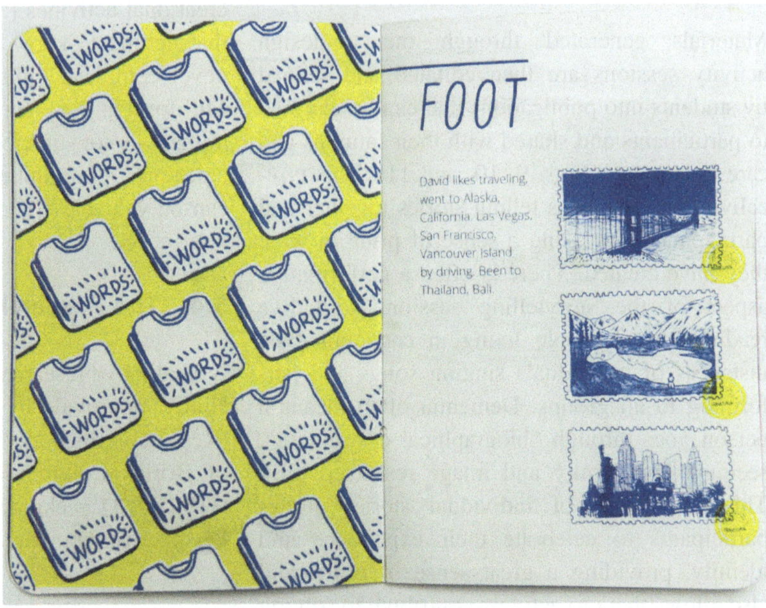

Fig. 10 Example of a group's finished publication

their feelings about the project and the perceived benefit for care home participants.

5.6 Findings

The findings of the research study highlighted many benefits to personhood for care home participants. Notably, staff remarked that they learned things about the people lving there that they did not know before and the final publications also allowed for staff outside of the program to learn more about the participants as well. By utilizing a group format that brought together multiple people living in care and students together, participants also learned things about each other and made new connections and friends within the home—addressing some of the

Fig. 11 Example of a group's finished publication

isolation challenges often facing individuals when they transition into a new care home.

Families also learned new things about their loved ones, and were able to have their stories documented in a publication they could keep. Unlike the 'Getting to Know Me' form, the information wasn't simply biographical, covering their lives before entering the home, with many of the publications containing new works. When staff were asked the impact, they believed the program to be having on care home participants one responded:

> Not everybody is going to remember that experience or what happened in the moment, but this document or publication is going to allow them to share with other people, so the lasting experience will go beyond just meeting in the group. Many of our residents have short-term memory loss, but have long-term memory intact. They are able to remember the stories, but they are going to forget they did these group sessions. Family members and fellow residents are going to be able to recall and share this publication with them and enjoy it with others and that keeps the whole emotional experience alive.

When asked how they think this program will benefit or impact care home participants, one responded:

> You also have the staff relationship. The staff will know more about the resident so it's a much easier way to connect. Much like a resident-to-resident. There is no more of a difference but to know someone more and provide meaningful care and honour their personhood and knowing that as staff that this is their person then it goes a long way to affect the way we provide care to them.

When asked how they think the program will impact their knowledge of the people living in care, one staff member responded:

> Absolutely. It will impact the knowledge of our residents… I really want to see how engaged everyone of our community is in this opportunity. It reminds us that we are in a village and everyone has a role and often the elders are the storytellers and they teach us history, they teach us lessons and they give us the experience we need to learn from. Yes we are an institution, yes we provide care in a formalized setting, through the government and all that, but at the end of the day, the care home is a village and it's giving it back to the elders and giving them back and saying you are the storytellers, the shamans and the wise men of the village and so we want to empower them…

5.7 Next Steps

Following four successful incarnations of the program at two different care homes, a How to Guide has been developed to allow other institutions to introduce the program within their curriculum (www.perspectivesprogram.ca). The program has also led to conversations about how to further enable these stories to be shared beyond the program itself—such as adding the publications to the care home library, sharing them during staff meetings and creating pub nights for story sharing.

6 Discussion

Dementia care is big business and growing. To ensure the effective delivery of future services, the current system of care needs immediate attention. There appears to be a gap between the wealth of research in the field of dementia care—and person-centred care more specifically, and the application of this knowledge at a grassroots level of care. Our case studies illustrate the importance of a person's life history and identity for people with dementia—both to their sense of self and in the creation of quality long-term care. Understanding a person's history, their likes and dislikes, their interests and sensitivities can have an immediate positive impact on care within all environments and can help to support the experience of transitioning into a new home, and a new community. The information gathered about each individual can inform the approach to their care and can be transferred between different care providers to maintain quality care at all stages and within all environments. The important challenge that our case studies address, however, is the development of creative and accessible methods for collecting and documenting these histories.

The Design + Storytelling framework is a way of better understanding people through storytelling [59]. The framework advocates the use of storytelling as a research methodology, as storytelling reveals more social context than a typical interview style of investigation. Storytelling is naturally a collaborative activity, with a story being created not just by the person telling the story, but also by the person listening to the story and responding to it. Stories expressed by people living in care during the activity sessions in Case Study 2 undergo a process of narrative analysis to identify the key themes that are important to the participants, and they then become the focus of further exploration and future activities.

The ReCollect system (Case Study 1) and Perspectives Program (Case Study 2) enables the residents to translate information into a permanent record through the power of storytelling. Going through this process in the early stages of dementia helps to reinforce the person's sense of self and identity [9] The activity brings families to a deeper level of understanding, and by asking specific questions, it begins to prepare them for the later stages and the potential move to permanent care. In the long-term, when the person with dementia begins to experience memory loss or they have difficulty communicating for themselves, this collection of valuable information provides starting points for conversation and interaction with others. Additionally, the process of reflection and reminiscence [60] may also provide a welcome distraction from the everyday worry of dealing with dementia. These storytelling approaches have a lot of potential as a better ways to get to know people living in care homes. We have shown just two ways in which this can be done, but there are likely to be many more.

Our findings reflect Kitwood's notion of person-centred care [9], where observations and personal accounts from carers indicate that this perspective, and the use of user-specific information, deserve priority. Non-pharmacological interventions, recognising individual needs and the benefits of person-centred care, are increasingly being used and trialled in care environments worldwide. However, many outdated work cultures do not give this perspective priority.

The challenges we see with these concepts that future iterations will need to address are threefold: (1) strategies to manage sensitivities around emotionally charged experiences and the recollection of potentially painful memories; (2) the timing of activities to coincide with early stages of the illness to accommodate communication and informed consent; and (3) managing the personal information once the person with dementia is no longer able to give informed consent, or indeed, after they have passed away. While we see a significant benefit in such information being able to communicate aspects of a person's life story to new and 'unknown' caregivers, challenges remain on maintaining truth and respect to the voice of the person. Further, in both cases, consideration for how these design approaches can be utilized and integrated by staff within the context of their duties is required.

7 Conclusion and Future Work

True understanding comes from conscious listening in conversations. Design methods provide accessible ways to encourage and engage older people and their caregivers in a process of reflection and reminiscence that brings a deeper level of understanding. Such methods also provide opportunities to explore respectful, creative and effective ways to collect and share valuable life stories and experiences so that they may contribute to genuine quality, personalised care for people with dementia. Future research and practice should consider a greater emphasis on the use of personal life history information in dementia care. A challenge we are currently facing is how to recontextualize the work we do for a post-pandemic landscape that restricts access to long-term care homes in a time when connection to people and community is more essential than ever.

References

1. World Health Organisation.: Dementia: a public health priority (2012). https://www.who.int/mental_health/publications/dementia_report_2012/en/
2. Alzheimer's Disease International.: www.alz.co.uk/. (2020)
3. Prince, M., Comas-Herrera, A., Knapp, M., Guerchet, M., Karagiannidou, M.: World Alzheimer report 2016: improving healthcare for people living with dementia: coverage, quality and costs now and in the future (2016). Retrieved from http://eprints.lse.ac.uk/67858/1/Comas-Herrera_World%20Alzheimer%20report_2016.pdf
4. Brown, L., Hansnata, E., La, H. A.: Economic cost of dementia in Australia. Alzheimer's Australia, Canberra (2017). Retrieved from https://www.dementia.org.au/sites/default/files/NATIONAL/documents/The-economic-cost-of-dementia-in-Australia-2016-to-2056.pdf
5. Alzheimer Society of Canada.: Dementia number in Canada (2020). https://alzheimer.ca/en/about-dementia/what-dementia/dementia-numbers-canada
6. Prince, M., Bryce, R., Ferri, C.: ADI—world Alzheimer report 2011 (2011). Retrieved from https://www.alzint.org/resource/world-alzheimer-report-2011/
7. Caddell, L. S., Clare, L.: Interventions supporting self and identity in people with dementia: a systematic review. Aging & Ment. Health **15**(7), 797–810. (2011, 2011/09/01). https://doi.org/10.1080/13607863.2011.575352
8. Edvardsson, D., Winblad, B., Sandman, P.: Person-centred care of people with severe Alzheimer's disease: current status and ways forward. Lancet Neurol. **7**(4), 362–367 (2008)
9. Kitwood, T. Dementia reconsidered: the person comes first. Open University Press. (1997a)
10. Kitwood, T.: The experience of dementia. Aging & Ment. Health **1**(1), 13–22 (1997). https://doi.org/10.1080/13607869757344
11. Dementia UK.: Life story template. http://www.dementiauk.org/information-support/life-story-work/. (2012)
12. Alzheimer's Society.: This is me leaflet. http://www.alzheimers.org.uk/site/scripts/download_info.php?fileID=849. (2012)
13. Innes, A., Manthorpe, J.: Developing theoretical understandings of dementia and their application to dementia care policy in the UK. Dement. **12**(6), 682–696 (2013)
14. Cox, H., Burns, I., Savage, S.: Multisensory environments for leisure: promoting well-being in nursing home residents with dementia. J. Gerontol. Nurs. **30**(2), 37 (2004)
15. Rasin, J., Kautz, D.D.: Knowing the resident with dementia: perspectives of assisted living facility caregivers. J. Gerontol. Nurs. **33**(9), 30 (2007)
16. Patel, B., Perera, M., Pendleton, J., Richman, A., Majumdar, B.: Psychosocial interventions for dementia: from evidence to practice. Adv. Psychiatr. Treat. **20**(5), 340–349 (2014). https://doi.org/10.1192/apt.bp.113.011957
17. Cohen-Mansfield, J.: Nonpharmacological interventions for persons with dementia. Alzheimer's Care Today **6**(2), 129 (2005)
18. Gräsel, E., Wiltfang, J., Kornhuber, J.: Non-drug therapies for dementia: an overview of the current situation with regard to proof of effectiveness. Dement. Geriatr. Cogn. Disord. **15**(3), 115–125 (2003)
19. Moos, I., Bjorn, A.: Use of the life story in the institutional care of people with dementia: A review of intervention studies. Ageing Soc. **26**(3), 431–454 (2006)
20. Travers, C., Bartlett, H.P.: Silver memories: implementation and evaluation of a unique radio program for older people. Aging & Ment. Health **15**(2), 169–

177. (2010, 2011/03/01). https://doi.org/10.1080/13607863.2010.508774
21. Gill, L., White, L., Cameron, I.D.: Interaction in community-based aged healthcare: perceptions of people with dementia. Dement. **10**(4), 539–554 (2011)
22. Popham, C., Orrell, M.: What matters for people with dementia in care homes? Aging & Ment. Health **16** (2), 181–188. (2011, 2012/03/01). https://doi.org/10.1080/13607863.2011.628972
23. Handley, M., Bunn, F., Goodman, C.: Dementia-friendly interventions to improve the care of people living with dementia admitted to hospitals: a realist review. BMJ Open **7**(7), e015257 (2017). https://doi.org/10.1136/bmjopen-2016-015257
24. Clissett, P., Porock, D., Harwood, R.H., Gladman, J. R.: The challenges of achieving person-centred care in acute hospitals: a qualitative study of people with dementia and their families. Int. J. Nurs. Stud. **50** (11), 1495–1503 (2013)
25. Bradford Dementia Group.: Dementia care mapping. Retrieved 12 April 2012 from http://www.brad.ac.uk/health/dementia/dcm/. (2012)
26. Chenoweth, L., King, M.T., Jeon, Y.H., Brodaty, H., Stein-Parbury, J., Norman, R., Haas, M., Luscombe, G.: Caring for Aged Dementia Care Resident Study (CADRES) of person-centred care, dementia-care mapping, and usual care in dementia: a cluster-randomised trial. Lancet Neurol. **8**(4), 317–325 (2009)
27. Edvardsson, D., Innes, A.: Measuring person-centered care: a critical comparative review of published tools. Gerontol. **50**(6), 834–846 (2010)
28. Burke Deborah, M., Mackay, D.G. Memory, language, and ageing philosophical transactions of the royal society B **352**, 1845–1856 (1997). https://doi.org/10.1098/rstb.1997.0170
29. Royal College of Speech and Language Therapists (RCSLT): Communicating Quality 3, 3rd edn. RCSLT, London (2006)
30. Banovic, S., Zunic, L.J., Sinanovic, O.: Communication difficulties as a result of dementia. Mater. Socio—Med. **30**(3), 221 (2018)
31. Kovach, S.S., Robinson, J.D.: The roommate relationship for the elderly nursing home resident. J. Soc. Pers. Relationships. **13**(4), 627–634 (1996). https://doi.org/10.1177/0265407596134009
32. Milte, R., Shulver, W., Killington, M., Bradley, C., Ratcliffe, J., Crotty, M.: Quality in residential care from the perspective of people living with dementia: the importance of personhood. Arch. Gerontol. Geriatr. **63**, 9–17 (2016)
33. Reuss, G.F., Dupuis, S.L., Whitfield, K.: Understanding the experience of moving a loved one to a long-term care facility. J. Gerontol. Soc. Work **46**(1), 17–46 (2005, 2005/11/22). https://doi.org/10.1300/J083v46n01_03
34. Heliker, D., Scholler-Jaquish, A.: Transition of new residents to long-term care: basing practice on residents' perspective. J. Gerontol. Nurs. **32**(9), 34 (2006)
35. Kenning, G.: Reciprocal design: inclusive design approaches for people with late stage dementia. Des. Health **2**(1), 142–162 (2018). Retrieved from https://doi.org/10.1080/24735132.2018.1453638
36. Sanders, E.: From user-centered to participatory design approaches. design and the social sciences. Boca Raton: CRC Press, 18–25 (2018). Retrieved from https://doi.org/10.1201/9780203301302
37. Alzheimer's Society.: (2020) https://www.alzheimers.org.uk/
38. Miles, M.B., Huberman, A.M., Saldana, J.: Qualitative data analysis: a sourcebook of new methods, 3rd edn. Sage Publications, London (2014)
39. Saldaña, J.: The coding manual for qualitative researchers: Sage [EBL version] (2009). Retrieved from http://www.qut.eblib.com.ezp01.library.qut.edu.au/patron
40. Froggatt, K.: Life and death in English nursing homes: sequestration or transition? Ageing Soc. **21** (3), 319–332 (2001)
41. Barnabe, A., Whitehead, V., Pilon, R., Arsenault-Lapierre, G., Chertkow, H.: Auto biographical memory in mild cognitive impairment and Alzheimer's disease: A comparison between the Levine and Kopelman interview methodologies. Hippocampus (2012)
42. Levine, B., Svoboda, E., Hay, J.F., Winocur, G., Moscovitch, M.: Aging and autobiographical memory: dissociating episodic from semantic retrieval. Psychol. Aging **17**(4), 677 (2002)
43. Moss, H., Cappelletti, M., De Mornay Davies, P., Jaldow, E., Kopelman, M.: Lost for words or loss of memories: autobiographical memory in semantic dementia. Brain Lang. **74**(3) (2000)
44. Thomsen, D.: An interviewer's guide to autobiographical memory. Tidsskrift for Kvalitativ Metodeudvikling **41**, 81–95 (2006)
45. Svoboda, E.M.: Autobiographical interview: age-related differences in episodic retrieval University of Toronto (2002)
46. Concordia University.: Sample life story interview questions (n.d.). http://storytelling.concordia.ca/oralhistory/resources/tips/Documents/Sample_Questions.pdf
47. Smithsonian Institution.: The smithsonian folklife and oral history interviewing guide. (2003) http://www.folklife.si.edu/resources/pdf/interviewingguide.pdf
48. Hannan, J., Raber, C., Ellis, E., Beyzaei, N., Levi, D., Phinney, A., Sakamoto, M.: Zeitgeist publication: a storytelling project with residents and design students. Des. Health (2019). Retrieved from https://doi.org/10.1080/24735132.2019.1596210
49. Katsuno, T.: Dementia from the inside: how people with early-stage dementia evaluate their quality of life. Ageing Soc. **25**(2), 197–214 (2005). Retrieved from https://doi.org/10.1017/S0144686X0400279X

50. Rodgers, P.: Co-designing with people living with dementia. CoDesign **14**(3), 188–202 (2018). Retrieved from https://doi.org/10.1080/15710882.2017.1282527
51. Hendriks, N., Slegers, K., Duysburgh, P.: Codesign with People Living with Cognitive or Sensory Impairments: A Case for Method Stories and Uniqueness. CoDesign **11**(1), 70–82 (2015). Retrieved from https://doi.org/10.1080/15710882.2015.1020316
52. Langley, J., Wolstenholme, D., Cooke, J.: 'Collective making' as knowledge mobilisation: the contribution of participatory design in the co-creation of knowledge in healthcare. BMC Health Services Research, 18 (2018). Retrieved from https://dx.doi.org/10.1186%2Fs12913-018-3397-y
53. Vikström, S., Josephsson, S., Stigsdotter-Neely, A., Nygard, L.: Engagement in activities: experiences of persons with dementia and their caregiving spouses. Dement. **7**(2), 251–270 (2008). Retrieved from https://doi.org/10.1177/2F1471301208091164
54. Redström, J.: RE: definitions of use. Des. Stud. **29**(4), 410–423 (2008). Retrieved from https://doi.org/10.1016/j.destud.2008.05.001
55. Branco, R,M., Quental, J. Ribeiro, Ó.: Personalised participation: an approach to involve people with dementia and their families in a participatory design project. CoDesign **13**(2), 127–143 (2017). Retrieved from https://doi.org/10.1080/15710882.2017.1310903
56. Hydén, L.: Storytelling in dementia. living with dementia: relations, responses and agency in everyday life, pp. 116–136. UK, Macmillan Education (2017)
57. Bury, M.: Chronic illness as biographical disruption. Sociol. Health & Illn. **4**(2), 167–182 (1982). Retrieved from https://doi.org/10.1111/1467-9566.ep11339939
58. Manzini, E.: Design, when everybody designs: an introduction to design for social innovation. MIT Press, New York (2015)
59. Galliers, J., Wilson, S., Roper, A., Cocks, N., Marshall, J., Muscroft, S, Pring, T.: Words are not enough: empowering people with aphasia in the design process. Proceedings of the 12th Participatory Design Conference: Research Papers, 1, 51–60 (2012). Retrieved from http://dx.doi.org/10.1145/2347635.2347643
60. Gausepohl, K., Winchester, T.W., Smith-Jackson, B., Kleiner, J. Arthur, J. A conceptual model for the role of storytelling in design: leveraging narrative inquiry in user-centered. Design. Health Technol. **6**(2), 125–136 (2016). Retrieved from https://doi.org/10.1007/s12553-015-0123-1

Exploring How a Multisensory Media Album Can Support Dementia Care Staff

Ward de Groot, Gail Kenning, Elise van den Hoven, and Berry Eggen

Abstract

For dementia care staff working in residential facilities, it may be challenging to interact on an interpersonal and multisensory level with residents. The progressive nature of dementia influences a person's physical and cognitive functioning. Research has shown that offering multisensory interactions for people with dementia can improve their behavior and mood. We are interested in gaining an understanding of how dementia care staff could be supported to engage with people with dementia. To explore this, a multisensory media album was designed to encourage care staff to interact with their residents on an interpersonal and multisensory level. Observations and interviews gave insight into how dementia care staff experienced the design, and how it helped them to interact with a variety of residents. We suggest that future design-researchers explore alternative means of multisensory engagement by utilizing photos, audio fragments, and textiles, as these showed potential to support dementia care staff to connect and communicate with residents.

Keywords

Dementia · Dementia care staff · Multisensory stimulation · Reminiscence · Meaningful associations · User-Centered design

W. de Groot · E. van den Hoven · B. Eggen
Department of Industrial Design, Eindhoven University of Technology, Eindhoven, The Netherlands

W. de Groot · G. Kenning (✉) · E. van den Hoven
University of Technology Sydney, Ultimo, Australia
e-mail: Gail.Kenning@unsw.edu.au

E. van den Hoven
University of Dundee, Nethergate, Dundee, Scotland, UK

G. Kenning
Ageing Futures Institute, fEEL (Felt Experience and Empathy Lab), University of New South Wales, Kensington, Australia

1 Introduction

Dementia care staff working in residential facilities are posed with care challenges due to the diverse and complex nature of dementia [1]. Dementia is a progressive disease which influences a person's physical and cognitive functioning. People with dementia are confronted with cognitive decline, affecting their ability to communicate with others. These communication difficulties are often experienced as frustrating for people with dementia, and can be distressing for care staff, which in turn can affect the staff's ability to identify the needs of a person with dementia [2].

R. Brankaert et al. (eds.), *Dementia Lab 2021: Supporting Ability Through Design*, Design For Inclusion 2, https://doi.org/10.1007/978-3-030-70293-9_4

To this day, there is no cure for dementia. As design researchers, it is therefore interesting to explore non-pharmacological alternatives to satisfy the needs of people living with dementia, as these can support alleviating distress [3], and improve the quality of life and well-being of people living with dementia [4]. Technology plays a role in complying to their needs, as it can enrich their daily life activities [5]. In addition to using assistive technologies, residential care facility staff can incorporate other strategies to engage with their residents on a personal level. Strategies may include but are not limited to using multisensory stimulation (MSS), using tactile elements, and organizing reminiscing activities.

2 Background

2.1 Multisensory Stimulation

MSS [6] is an approach aiming to stimulate the senses in a gentle manner without the need for other higher cognitive processes such as learning [7, 8]. MSS allows caregivers to provide person-centered care [9] to people with dementia, as it requires a resident-oriented attitude, knowledge, and skills. Moreover, when MSS is introduced in daily care activities, it can positively impact residents' behaviors and mood [10, 11], reduce anxiety [12], and increase resident-staff interactions [13]. Furthermore, the utilization of tactile elements within the dementia care space can offer opportunities of sensorial exploration and experience, fulfilling one's preferences and psychological needs, including comfort, security, identity, occupation, and pleasure [14].

2.2 Reminiscence in Care Practices

Reminiscence activities are popular for engaging with people with dementia [15]. Individual reminiscence activities have reportedly been successful for people with dementia, benefiting mood, well-being and aspects of cognitive functioning [16]. Moreover, Morgan and Woods suggest that tangible outcomes of reminiscence work, such as life story books and memory boxes, may prolong the effects of reminiscence and act as a maintenance tool for people with dementia living in care homes [17].

One way to stimulate reminiscence is by using music [18]. Music also brings about other benefits, as it has been shown to facilitate social interactions [19], relieve stress and boredom [20], encourage physical movement [21], and act as a stimulus for exploring and maintaining identity [22]. Additional to music, sounds of everyday life are another category of sounds that human beings can perceive. These sounds are important for us, as they leverage communication, provide information, build expectations, and cue behaviors [23, 24]. Everyday sounds have also shown to affect the mood and behavior of people with severe or profound intellectual disabilities in a positive way [25].

The majority of studies in residential dementia care focus on the people with dementia, rather than focusing on care staff working in residential care facilities. Dementia care staff often deal with high working pressure, and forms of aggression [26]. A recent study revealed indications that a film installation allowed dementia caregivers to connect with people with dementia through made associations followed from the contents in the film [27]. Moreover, Treadaway et al. [28] suggest that visual and tactile sensory cues can prompt deep memories and associations, even though the triggers for these are highly personal and related to individual preferences and experiences.

Drawing on reminiscing practices, we argue that designing a combination of MSS elements into an object that prompts familiarity through associations can support dementia care staff, improve the quality of the care they provide, and can improve their working environment. Therefore, we aim to design a tangible, psychosocial intervention, which can support care staff to engage, connect, and communicate with people with dementia.

In this paper, we present the design of a multisensory media album, which focused on assisting residential care staff members at Leigh Place Aged Care, Sydney, Australia to interact on a personal and multisensorial level with their residents using a combination of visual, auditory, and tactile cues. These cues aim to prompt recognition for both care staff members and the residents, so they can build upon each other's prompt responses to create a connection.

3 Designing the Multisensory Media Album

The design is the result of a thorough contextual exploration, which consisted of an observation of care staff and a creative workshop with care staff. Here, we offer insights in how dementia care staff experienced using the design during their shifts, and how it helped them to interact with a variety of residents.

3.1 Methodology

3.1.1 Observation and Creative Workshop

Prior to the design of the multisensory media album, dementia care staff were invited to take part in an observation and a creative workshop.

An observation was done to obtain initial insights in how care staff interact with people with dementia on a regular basis. By making field notes and through informal conversations with care staff during the observation, an understanding was gained of the care staff's routine activities, their overall experiences during their shifts, their ways of engagement with residents, and opportunities for engagement were identified.

The creative workshop aimed to acquire further insights in how care staff might engage with people with dementia through physical or digital media. Five care staff members made low-fidelity prototypes, which were a representation of what they deemed suitable forms of engagement for people with dementia. To elaborate, a personalized book was made, while other prototypes included a television, maracas, a shaker, dumbbells, binoculars, and a gardening area. These prototypes indicated different engagement opportunities by for example promoting physical activity, and engaging in music activities. These prototypes, combined with the observational insights, inspired the design of the multisensory media album.

The observation and creative workshop were approved by the Human Research Ethics Committee of the University of Technology Sydney (ETH19-4209) and by the Ethical Review Board of the University of Technology Eindhoven (ERB2019ID9). All research activities described within this paper were approved by the managing staff of Leigh Place in Roselands, Sydney.

The field study and evaluation described in Sect. 4 were approved by the internal Materialising Memories Committee of the University of Technology Sydney and by the Ethical Review Board of the University of Technology Eindhoven (ERB2020ID1).

3.2 Design Rationale

As mentioned above, the prototypes made inspired the design of the multisensory media album. For example, one participant made a prototype of a reading book. Other prototypes included a television and a tin can phone. These prototypes allow for different kinds of interaction. For instance, the personal reading book and television can be utilized to present forms of physical and digital media, and the tin can phone can be used to encourage social interactions amongst residents, or between care staff and residents.

As suggested by Jakob and Collier [14], caregivers will be able to respond to individual preferences, needs, and abilities by offering a multisensory experience. Additionally, they state that a person with dementia might find it easier to relate to environments and occupation combining sensorial input on multiple levels, as the world we live in is also highly multisensory. For the resident, this means that the design should aim to provide more than one way to connect to, and enjoy their surroundings. By basing the multi-sensory media album on a photo book style object, this can be facilitated. A photo book presents a personal collection of memories that is often shared with others, promoting and suggesting a reminiscence activity.

For the pictures, we chose to only include generic photographs as opposed to personal, as people with dementia were observed to produce more detailed and emotional stories of personal significance when using generic photographs [29]. However, the photographs used should try to reflect the interests and abilities of the resident, as these will be more likely to motivate them to engage in the activity [30, 31]. Through the themes, we tried to represent the community of Leigh Place, addressing a mix of national and international topics of interest, which would prompt recognition for the 14 residents. For example, we chose to incorporate themes of an Australian children's show on page 1, well-known Australian beaches on page 4, and historical poetry on page 7 (Table 1). The other pages represent more internationally known themes, with generic photos and sounds. The textiles were chosen to fit the pictures. For example, for the theme Christmas, white wool is used, as it could imply feelings of warmth (Fig. 1).

By designing the album in this way, it allows for reciprocal interaction between care staff and residents. For example, the care staff can make associations with, and elaborate on the pictures and audio fragments, and can encourage a resident to touch the textiles. In return, the resident's senses are stimulated and could remember things of the past.

3.3 Design Description

The prototypes made during the creative workshop presented a variety of ways to engage with people with dementia, which were therefore incorporated in the design. Our design utilizes music, pictures, and textiles for care staff to interact with residents. An off-the-shelf photo album was modified to feature a conductive touch pad on each page of the album. Each touch

Table 1 Themes, audio, and textiles used in the multisensory media album

Page	Theme of pictures	Audio	Textile
1	Australian children's show	Intro song of the show 'Skippy the bush kangaroo'	Brown linen with sandy texture
2	Birthday	Song 'Happy birthday to you', karaoke version	White linen with soft texture
3	Wedding	Song 'Let's twist again' by Chubby Checker	Pink felt
4	Beach	Sound of waves crashing, recorded at Bondi Beach	Pink polyester with flower pattern
5	Christmas	Song 'We wish you a merry Christmas', karaoke version	White wool
6	Historical poetry	Narration of 'Scots of the Riverina', poem by Henry Lawson	Brown linen with sandy texture
7	Nature	Composition of 'Eine kleine Nachtmusik' by Wolfgang Amadeus Mozart	Pink felt

Fig. 1 The multisensory media album's pages (1 to 7 from left to right) as presented to the residential care staff. The different themes on each page are accompanied by textile matching the theme. Audio fragments related to the theme are triggered when the conductive pads are touched

pad triggers an audio fragment related to the theme brought forth by the pictures on that page (Fig. 1). Whenever a sound or song was triggered, it could be stopped either by touching another pad, or by touching one of the sensors on the back of the album. Furthermore, the pages of the album are covered with different textiles for the resident to engage with (Table 1).

4 Engaging with the Multisensory Media Album

4.1 Study Setup

A user-centered design approach [32] was applied to explore how a multisensory media album could assist caregivers to engage on a multisensory and personal level with people with dementia. All participants described in this study took part on a voluntary basis and provided consent, which was accompanied by an information sheet. Any pictures taken were processed anonymously, and emphasized on interactions with the multisensory media album, rather than on any individual.

Five care staff members participated in the field study. We were interested in how, when, and why care staff would interact with residents using the multisensory media album. We were also interested whether the amount of experience in the field and their current role at the care facility would influence the degree in which the multisensory media album would be able to support the care staff. Table 2 shows a demographic overview of the participants.

Participants were introduced to the multisensory media album, and they were shown how it works through a demo. The participants were instructed to focus on their regular duties during their shift, and were told that they were free to use the multisensory media album whenever they wanted. Observations were done during this time, which took up to two hours per participant.

During each observation, notes were made to try and capture the ways of how care staff engaged with the residents through the multisensory media album. The notes also gave an insight in when the album was used. Several caregiver-resident interactions were observed. After each observation, a semi-structured interview [33] was held with each participant. The interviews were audio recorded for post-event analysis. The questions focused on the care staff's general responsibilities, their experiences of using the multisensory media album, and how they thought the residents experienced the multisensory media album.

A thematic analysis [34] was conducted using an inductive approach to reveal reoccurring themes within the observational field notes and semi-structured interviews. In total, five themes were identified.

4.2 Care Staff's Duties and Responsibilities

As seen in Table 2, the participants had varying occupations at the care facility. A clear distinction between Recreational Activities Officers (RAOs) (P1, P3) and regular caregivers (P2, P4,

Table 2 Demographical data of participants

Participant	Male/Female	Occupation	Time employed in current position	Total time in aged care
P1	Female	Recreational Activities Officer	1 year	2 years
P2	Female	Regular caregiver	14 years	14 years, 8 months
P3	Female	Recreational Activities Officer	10 years	19 years
P4	Female	Regular caregiver	5 years	10 years
P5	Female	Regular caregiver	22 years	22 years

P5) was evident from the observations and interviews. The RAOs' duties and responsibilities were described as to "*encourage the residents to come to activities, and also engage for socialization and to make the residents feel like home and happy, and be aware if someone is in need of anything to or to find out what they need, and to fill that in for them*" [P1]. The regular caregivers are responsible for routine care activities as identified in the initial observation described in Sect. 3.1. When asked what regular caregivers found the most important thing they did, answers varied from "*...make them happy...*" [P2], and "*Respect their choices and dignity.*" [P5] to "*Everything*" [P4].

Due to these differences in care activities, RAOs were observed to engage with a higher number of residents, and overall spend more time per resident with the multisensory media album. P4 also mentioned that the RAO "*...would probably benefit from (the multisensory media album) everyday, because she's got time. She does one on one sessions with the residents, so she might spend, I don't know how long, she spends half an hour or whatever it is and she could take that and use it.*"

4.3 Interactions Between Care Staff and Residents

The design was observed to provide a way for care staff to engage with residents. Care staff initiated an interaction by introducing the album to several residents: "*I want to show you something, it's very interesting*" [P3], and "*Can I show you a book I have?*" [P1]. In total, 16 interactions were initiated.

Also, care staff were observed to try and encourage residents to interact with the album by asking questions and giving instructions: "*How does that feel like?*" [P1] (Fig. 2), "*You have to dance! Do you like it?*" [P2], "*Touch it here*" [P3].

The interactions with the album were observed to offer an enjoyable experience for both care staff and residents. For example, P2 was observed holding a resident's hand and asking him: "*You and me, are we going to marry?*" on the wedding themed page. This resulted in them laughing together. Moreover, P1 noted the following on the behavior of several residents: "*...they were tapping, uhm. I could see them recalling and thinking back of memories, it was bringing back a memory that was making them happy. And they were singing. [...] They were relating the texture to the pictures.*" This indicates that the care staff had the impression that residents had an understanding of the represented themes in the album, due to their behavioral changes, which was also evident from P3's statement: "*...you can see the reaction from some of them, when they're touching and the music catching their attention so they're looking, and also if you have questions they're looking, and some of them they can identify the objects and the pictures.*"

Overall, engagement was sought successfully through the album. However, there were two occasions in which two separate residents did not

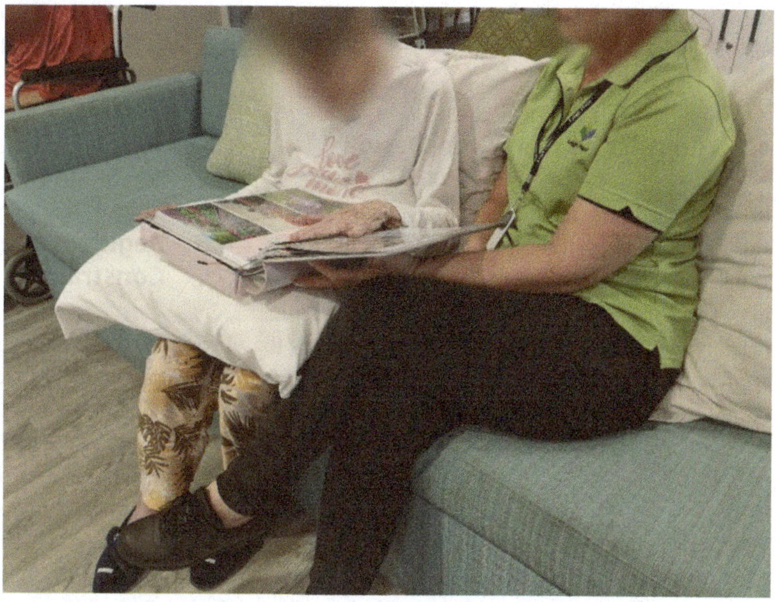

Fig. 2 P1 encouraged the resident to touch the textile

seem to be interested in any interaction with the album. The care staff elaborated on these moments in the conducted interviews: "*At the moment he's scratching his leg so he's itchy. So that is distracting him at the moment*" [P1], and "*He was not in the mood, he just wanted to go. Just walk.*" [P5]. The resident that showed no interest to P1, was observed to enjoy P2's and P3's engagement when introduced to the album, supporting P1's statement.

4.4 Care Staff's Thoughts on the Design

During the interviews, care staff expressed their thoughts on the multisensory media album. They reported that it supported them during their care activities: "*Yes! For the one on one engagement, individual engagement with residents, I find that this would be very beneficial, because it has a memory recall for them, it's helping them to engage with you, and I see the enjoyment in their face. And all residents respond differently: some do the texture things, some will sing along, some will be tapping, and you can see their face light up, and that's what we want.*" [P1], and "*Yes, yes! […] Because it's less stress (for) both of us, especially for the residents*" [P2].

P3 commented on how the multisensorial cues of the album could distract residents for longer periods of time: "*It's good, especially for people with dementia. Because they can have at the same time the music and the pictures, that is coming together and it can be good for them, that they can, like I said, catch their attention, […] If you show them a picture or a book, they can have a look for a little time before they get bored, and with this project I think so they can stay for a little while observing, and listening, so they can engage for a longer time.*"

Moreover, P1 made a remark on how she thought the design helped the residents: "*Things to engage a resident, that's the number one thing that really helps you can see them coming out of their shell and it helps… I think it helps them to focus on something, and it's sort of pulling them out of the place that they're in, their own world. And it's bringing them happiness, and that's all we want.*"

Furthermore, participants shared some design considerations that could be used for further development of the album: "*And something that you could do more: shows from the past with pictures to help them recall, or even old ads that were on TV.*" [P1], and "*It has to be happy music*" [P2].

4.5 Context of Using the Design

In addition to P1 stating that the design would support her during one on one engagement sessions, other insights were gained on the context in which the design was used. For example, participants deliberately chose which residents to approach with the album: "*Yes, it depends which residents, [...] we already know the character or the impression on their face, yes, it's like we have to have a look to them, and choose to offer them*" [P2].

P1 also mentioned reasons for not engaging with residents: "*...they might be tired. [...] One resident might want me to sit down with them, they might want me to take them for a walk. Another resident might just want to sit down and talk to me.*" However, P1 stated that she would try to use alternative ways to engage with residents instead: "*So, I just gage by: I'll bring it out, if they're not interested in that I just do something else.*"

Residents were often accompanied by another person whenever the album was shared with them. In most cases, this was the participant themselves. However, P2 was observed sharing the multisensory media album with family members of a resident. After this moment, P2 went back to the kitchen area to continue preparing food for lunch. In the interview, P2 noted that "*they enjoyed with their family and especially [resident's name], he was impressed, they were listening to the music, and pressing the buttons with their family. And the granddaughter she was showing it to him. And when I asked to them, they said yes, we love it too, it's a very nice book, and also, in his face, he was showing happiness.*" P2 noted that it was a good thing to be able to share the album with family members whenever she did not have the time to engage with the resident.

There were also two instances in which a resident was left unaccompanied with the album for a short time. During the observation of P1, the resident browsed through the pages and stroked the textile, and during the observation of P3, the resident browsed through the pages of the album until P3 returned.

4.6 Regular Caregivers' Time Constraints

From the regular caregivers, P2 was the only one who managed to use the multisensory media album during the observation with more than one resident. P4 only had a short interaction with a resident while being instructed, and P5 tried to engage with a resident, but was unsuccessful as described in Sect. 4.2. All regular caregivers brought forward that they had too little time during their shift to spend a reasonable amount of time with the design: "*Well, if you have time, and you're sitting down, you can take that and give it to the residents and listen to them. But this afternoon, like you could see, there is no time where I could try it*" [P5], and "*But sometimes the staff, no person has time. I like to do many things, but the kitchen, the cleaning, the everything, we have to do that*" [P2].

The regular caregivers also communicated that they would appreciate it if they would be able to spend more time to engage with the residents: "*I would love to interact more with the residents, but we don't have the time*" [P5].

Regardless of the regular caregivers' limited time to engage with residents through the design, they were observed to interact with residents during their regular care activities, which was also evident from the observations. For example, both P2 and P5 were talking to the residents while doing the dishes, and P4 was talking with the residents while she was administering medication to the residents. P4 described why it was impossible for her to use the album during these activities: "*Definitely not at that time, no, we couldn't bring it in during those times, because we need full concentration at medication time.*"

5 Discussion

Overall, the multisensory media album was a helpful tool for care staff to engage with residents. It offered moments of joy for both care staff and residents, and mediated communication between them. This was done by care staff who were encouraging residents to interact with the visual, auditory, and tactile cues as featured on the multisensory media album. The interactions between care staff and residents were enriched by making meaningful associations with contents presented by the multisensory media album. This allowed residents to remember things of the past, and their senses to be stimulated.

As revealed in Sect. 4.2, a clear difference between the responsibilities of the regular caregivers and the RAOs was observed in our study sample. After the first research phase, it was thought that regular caregivers would be able to pick up a design as presented, and share it with residents in between the moments of regular care activities. Yet, this was not apparent from our observations in the second research phase. Regular caregivers were observed to be more task focused, leaving less to no room for incorporating a design to engage with residents. Regular caregivers did express their wishes to engage more with their residents, and were observed to compensate this by approaching residents with a combination of verbal and physical communication during their regular care activities. One time, P2 was observed to give the album to family members of a resident, so they could interact with their loved one, using the album. During this time, P2 went back to carry out her care duties. The observed actions of P2, P4, and P5 resonate with the statements of regular caregivers feeling they do not have the time to utilize a design as such during their shifts. The differences among the care staff's responsibilities therefore contributed to when and in which instances the multisensory media album was used.

These results suggest that the multisensory media album is more suitable for care staff who have similar responsibilities to the RAOs as described in this study. The RAOs' responsibilities compared to those of regular caregivers, make RAOs more flexible in creating moments of engagement with the residents. They were observed to engage with a higher number of residents, and overall spend more time per resident with the multisensory media album. However, it must be noted that the RAOs communicate and interact with residents as part of the job, while the job of the regular caregivers, as abovementioned, is more task oriented.

The amount of years of employment in aged care did not seem to contribute to the degree in which care staff was supported by using the multisensory media album. We assumed that the more years of experience our participants had working in aged care would have an influence on their adaptability when introduced to our design. Though, this was not evident from our sample. Rather, their occupation and responsibilities seemed to be the most influential factors.

Although the multisensory media album showed promising results, this was not the case in all instances. It is important to note that within the dementia care space, each and every individual has different desires and needs, at different points in time. By introducing the multisensory media album to care staff, we were able to offer them an alternative way to support them to engage with people with dementia.

Even though the sample size of this study was low, our findings showed promising results for future implications. Due to time limitations, no pilot test was conducted. Instead, care staff were approached for input for the design prior to deployment. Adjustments to the design will likely be beneficial for care staff to provide residents with more effective and longer periods of engagement. To achieve this, the album could include additional pages, featuring personal themes of interest, or themes that would trigger the memories of residents. Moreover, we support the suggestions made by P1 and P2 to incorporate TV shows and advertisements, and utilize joyful music.

Offering a MSS activity takes time, which was considered valuable throughout the study. Utilizing MSS during care practices may result in a longer attention span for alike activities, whilst stimulating cognitive processes for people with dementia by allowing them to make associations through photos and audio fragments. For these engagement strategies to be successful within residential care facilities, design-researchers should aim to make MSS activities more accessible, by for example incorporating these in the direct environment of care staff and people with dementia.

6 Conclusion

This paper has provided explorative insights in how a multisensory media album can support care staff to engage with people with dementia. The design provided dementia care staff with a multisensorial activity for their residents, which was observed to be enjoyed by both. By presenting care staff with an alternative way to interact with residents, they can be assisted during their care practices. When designing for people with dementia and their caregivers in care facilities, we should be aware of the organizational structure, as this can influence the degree to which design is used and experienced. With this work, we aim to inspire other design-researchers to develop assistive technologies supporting combinations of multisensory engagement, as these have shown to be beneficial for care staff and people with dementia.

Acknowledgements We would like to thank the staff at Leigh Place for their cooperation and facilitating this study. This research was supported by STW VIDI grant number 016.128.303 of The Netherlands Organization for Scientific Research (NWO), awarded to Elise van den Hoven.

References

1. Chang, E., et al.: Challenges for professional care of advanced dementia. Int. J. Nurs. Pract. **15**(1), 41–47 (2009)
2. Jootun, D., McGhee, G.: Effective communication with people who have dementia. Nurs. Stand. **25**(25), 40–46; quiz 47 (Feb. 2011)
3. Mitchell, G., Agnelli, J.: Non-pharmacological approaches to alleviate distress in dementia care. Nurs. Stand. **30**(13), 38–44 (2015)
4. Douglas, S., James, I., Ballard, C.: Non-pharmacological interventions in dementia. Adv. Psychiatr. Treat. **10**(3), 171–179 (2004)
5. Cahill, S., Macijauskiene, J., Nygård, A.-M., Faulkner, J., Hagen, I.: Technology in dementia care. Technol. Disabil. **19**(2–3), 55–60 (2007)
6. Hulsegge, J., Verheul, A.: Snoezelen: another world: a practical book of sensory experience environments for the mentally handicapped. Rompa, Chesterfield (1987)
7. Van Weert, J.C.M., Kerkstra, A., Van Dulmen, A.M., Bensing, J.M., Peter, J.G., Ribbe, M.W.: The implementation of snoezelen in psychogeriatric care: an evaluation through the eyes of caregivers. Int. J. Nurs. Stud. **41**(4), 397–409 (2004)
8. Van Weert, J.C.M., Van Dulmen, A.M., Bensing, J. M.: What factors affect caregiver communication in psychogeriatric care? In: Visser, A.M. (ed.) Alzheimer's Disease: New Research, pp. 87–117. Nova Biomedical Books, New York (2008)
9. Kitwood, T.M.: Dementia reconsidered: the person comes first. Open University Press (1997)
10. van Weert, J.C.M., van Dulmen, A.M., Spreeuwenberg, P.M.M., Ribbe, M.W., Bensing, J.M.: Behavioral and mood effects of snoezelen integrated into 24-hour dementia care. J. Am. Geriatr. Soc. **53**(1), 24–33 (Jan. 2005)
11. van Weert, J.C.M., Janssen, B.M., van Dulmen, A. M., Spreeuwenberg, P.M.M., Bensing, J.M., Ribbe, M.W.: Nursing assistants' behaviour during morning care: effects of the implementation of snoezelen, integrated in 24-hour dementia care. J. Adv. Nurs. **53** (6), 656–668 (Mar. 2006)
12. Finnema, E. et al.: The effect of integrated emotion-oriented care versus usual care on elderly persons with dementia in the nursing home and on nursing assistants: a randomized clinical trial. Int. J. Geriatr. Psychiatry **20**(4), 330–343 (Apr. 2005)
13. van Weert, J.C.M., van Dulmen, A.M., Spreeuwenberg, P.M.M., Ribbe, M.W., Bensing, J.M.: Effects of snoezelen, integrated in 24 h dementia care, on nurse-patient communication during morning care. Patient Educ. Couns. **58**(3), 312–326 (Sep. 2005)
14. Jakob, A., Collier, L.: Sensory design for dementia care—the benefits of textiles. J. Text. Des. Res. Pract. **5**(2), 232–250 (2017)
15. Woods, B., Spector, A., Jones, C., Orrell, M., Davies, S.: Reminiscence therapy for dementia. Cochrane database Syst. Rev. (2), p. CD001120 (Apr. 2005)
16. Subramaniam, P., Woods, B.: The impact of individual reminiscence therapy for people with dementia: systematic review. Expert Rev. Neurother. **12**(5), 545–555 (2012)
17. Morgan, S., Woods, R.T.: Life review with people with dementia in care homes: a preliminary randomized controlled trial. Nonpharmacol. Ther. Dement. **1** (1), 43–60 (2010)

18. Baird, A., and Samson, S.: Memory for music in Alzheimer's disease: unforgettable? Neuropsychol. Rev. **19**(1), 85–101 (Mar. 2009)
19. Favilla, S., Pedell, S.: Touch screen ensemble music: collaborative interaction for older people with dementia. In: Proceedings of the 25th Australian Computer-Human Interaction Conference: Augmentation, Application, Innovation, Collaboration, pp. 481–484 (2013)
20. Raglio, A., Filippi, S., Bellandi, D., Stramba-Badiale, M.: Global music approach to persons with dementia: evidence and practice. Clin. Interv. Aging **9**, 1669–1676 (2014)
21. Lesaffre, M., Moens, B., Desmet, F.: Monitoring music and movement interaction in people with dementia. In: Lesaffre, M., Maes, P.-J., Leman, M. (eds.) The Routledge Companion to embodied music interaction, pp. 294–303. Routledge (2017)
22. Baird, A., Thompson, W.F.: The impact of music on the self in dementia. J. Alzheimers Dis. **61**(3), 827–841 (2018)
23. Bakker, S., van den Hoven, E., Eggen, B.: Knowing by ear: leveraging human attention abilities in interaction design. J. Multimodal User Interfaces **5** (3–4), 197–209 (2012)
24. Eggen, B.: Interactive soundscapes of the future everyday life. In: Bakker, S., Hausen, D., Selker, T. (eds.) Peripheral Interaction, pp. 239–251. Springer International Publishing, 2016
25. van den Bosch, K.A., Andringa, T.C., Peterson, W., Ruijssenaars, W.A.J.J.M., Vlaskamp, C.: A comparison of natural and non-natural soundscapes on people with severe or profound intellectual and multiple disabilities. J. Intellect. Dev. Disabil. **42**(3), 301–307 (2017)
26. van der Mooren, F., Gielen, W.: Werknemersenquête zorg en welzijn. Uitkomsten 1e meting 2019 (2019)
27. Bennett, J., Froggett, L., Kenning, G., Manley, J., Muller, L.: Memory loss and scenic experience: an arts based investigation. Forum Qual. Sozialforsch. **20**(1), 2019
28. Treadaway, C., Kenning, G., Prytherch, D., Fennell, J.: LAUGH: Designing to enhance positive emotion for people living with dementia. Proc.—D E 2016 10th Int. Conf. Des. Emot.—Celebr. Contempl. September, 37–44 (2016)
29. Astell, A.J., Ellis, M.P., Alm, N., Dye, R., Gowans, G.: Stimulating people with dementia to reminisce using personal and generic photographs. Int. J. Comput. Healthc. **1**(2), 177 (2010)
30. Bowlby, M.C.: Therapeutic activities with persons disabled by Alzheimer's disease and related disorders. Aspen Publishers, Gaithersburg, MD, US (1993)
31. Vozzella, S.: Sensory stimulation in dementia care—why it is important and how to implement it. Top. Geriatr. Rehabil. **23**(2), 102–113 (2007)
32. Abras, C., Maloney-Krichmar, D., Preece, J.: User-centered design. User-Centered Des., 445–456 (2004)
33. Longhurst, R.: Semi-structured interviews and focus groups. In: Key Methods in Geography, p. 43 (2010)
34. Virginia, V. Braun, Clarke: Using thematic analysis in psychology. J. Chem. Inf. Model **53**(9), 1689–1699 (2013)

Your Moments: Co-designing a Personalized Audio Player

Marjolein den Haan, Nicole van Essen, Rens Brankaert, and Yuan Lu

Abstract

People with memory challenges may experience a sense of losing themselves or 'unbecoming a self', preserving their personhood ensures increased quality of life. This paper explores a *design for one* approach to create a personal design tailored to our participants Adele, Bruce, and Chris. In collaboration with these three participants, an audio player was designed to provide access to meaningful audio files. The audio can vary from poems, music, messages from others, books, to personal messages from people to themselves serving as memories. In addition, we reflected on the directness and honesty of one of our participants in the design process, and our awareness that participants may not always be as open about their experience as Bruce was in this case. At last, we discuss the stakeholders who facilitated the use of 'Your moments', and we reflect on what this could look like in the future.

Keywords

Co-design · Dementia · People with memory challenges · Leisure time

1 Introduction

In our communities, people live with all kinds of memory challenges. Parkinson's disease, Lewy Body dementia, vascular disease, sleep apnea, kidney failure, delirium, depression and medication can cause challenges with cognition [1]. Some of these are progressive, and in the case of dementia there is no cure available [2]. In such a situation, design can help and add increased quality to their current lives [3]. This could for example be technology that facilitates meaningful interaction between people with dementia and their carers [4]. In dementia and other forms of cognitive impairment, people experience a decline in cognitive functioning and memory challenges that impact daily living. For these chronic conditions, people are often labelled as a homogenous group, yet they rather have very diverse and individual characteristics and needs [5]. As designers, we are interested to design for these individual wants and needs, and empower

M. den Haan (✉) · N. van Essen · R. Brankaert · Y. Lu
Department of Industrial Design, Eindhoven University of Technology, Eindhoven, The Netherlands
e-mail: m.c.wintermans@tue.nl

R. Brankaert
School of Allied Health Professions, Fontys University of Applied Sciences, Eindhoven, The Netherlands

R. Brankaert et al. (eds.), *Dementia Lab 2021: Supporting Ability Through Design*, Design For Inclusion 2, https://doi.org/10.1007/978-3-030-70293-9_5

them in the design process as well as their daily lives [6–8].

People with cognitive challenges may experience a sense of losing themselves or 'unbecoming a self' [9]. This may also be experienced by people with a brain tumour [10]. However, there is a need, according to Kitwood (11), for '*authenticity in preserving and respecting the personhood of the individual*' [6] when engaging people with dementia [12]. Research shows that one approach to strengthen personhood is using sound and music [13–15]. Everyday sounds were found to trigger emotional responses and encourage social engagement in people with dementia [13]. Furthermore, music allows people to strengthen their identity and reconnect them with their memories [15]. Treadaway [14] continues this work through personalized reminiscence to facilitate positive remembering through a selection of favourite music [14].

In design, it is essential to understand the perspective of the person with dementia to be able to create technologies that are suitable [16]. In design empathy is essential to understand the perspectives of different people [17]. Relevant work in the field of technology incorporate a person-centred perspective and bring people with dementia to the centre of the design process [18–20]. For example, Wallace et al. [18] show through the co-design of personal jewellery how design can aid in foregrounding personhood [18]. Lindsay et al. [19] reveal how empathic relationship between participants and designers lead to personally tailored designs [19], and Astell et al. [20] focus on learning new and old technology to create bespoke interventions [20].

Co-design with participants stimulates the creativity of both designers and participants and facilitates collaboration, this relationship is important [21]. In this study one designer collaborates closely with three participants, two who have dementia and one who had a brain tumour. The designer pursues to create an individualized and personal design solution for them. To create personal designs for people with memory challenges, the co-design approach needs to be adjusted to be both suitable for this group, and to allow for diverse and individual input. In this work we build further on Wilkinson et al. [22], by executing a design process that is inspired by the *designing for one* process. In the designing for one process the researcher propose to create design proposals to suit the individual's preferences, abilities, and context of participants because these are inherently related to how they cope with their situation [22]. By focussing on this individual situation and we can learn how dementia impacts people's lives on a larger scale [22]. In our process we conducted several co-creation sessions with the participants to understand their personal perspective [23]. In our approach we let the designer conduct this process with more individuals, so the differences and similarities emerge and bring the approach further.

Over the course of the highly individualized design process with the three participants, we created an audio player called 'Your moments', that caters for various individual needs and a diverse range of participants.

2 Approach and Results

This section consists of five parts: (1) participants, (2) a brief introduction of the process, (3) the final design, (4) results of phase 1: two designing for one processes, and (5) results of phase 2: exploring personalisation with a new participant. In this section approach and results are merged since they build further on each other and are intertwined.

2.1 Participants

In this section we elaborate on the inclusion of our participants in this study. For this we collaborated with a professional caregiver from care organisation Archipel Landrijt in Eindhoven, who recruited our participants. She understood our intentions and selected participants that were interested and able to consent on their own behalf. Written consent was given by the participants to process their data anonymously. Naturally, participants could withdraw from their participation in the research at any given moment. Additionally, the professional caregiver was closely involved in the project to support the participants. Therefore, she debriefed every session with the participants, and also met with the designer (second author) regularly.

The three participants were involved both during a day program organised by the care organisation and at their homes. Although the group is small, they had very different experiences in their life, which was reflected in their current abilities and interests. We call them by their pseudonyms Adele, Bruce, and Chris.

57-year old Adele was diagnosed with Alzheimer's disease and a motivated person with a strong opinion, who was keen on learning about things. She loved to spend time on being social, cooking, travelling, and making music. Adele enjoyed listening to music and graduated from music school. She was diagnosed with dementia at a young age and experienced difficulties adapting to her new situation. However, she also loved learning new things and was not afraid of a new challenge. Through the project she shared her expertise on aesthetics as she used to work as a beautician and appreciated a good appearance.

89-year old Bruce was diagnosed with early-stage dementia and a proud, social, and perfectionist man. He loved to spend time with his family and four lifelong friends, and he was a great storyteller. Bruce did not enjoy music as much anymore because his hearing decreased, and he started to hear different tones in music. Although, he had a difficult time trusting people around him he does not know, he still enjoyed being creative, he used to create sculptures from wood and was very interested in the making part of our design. After three sessions Bruce indicated he wanted to stop partaking in the study, but felt very proud that he had contributed, we will come back to this later.

Later in the process, 75-year old Chris was involved, who was diagnosed with a brain tumour. He was a positive, kind-hearted person who loved cooking, music and spending time with his family. He was very passionate about the catering industry in which he was involved since he was six years old, and he liked sharing his experiences from that period in time. Although cooking had become challenging due to his decreased sight, he—together with his wife—stayed positive and saw this project as a new and interesting challenge.

2.2 A Brief Introduction of the Process

The design process took place over several sessions between February and June 2019; these were audio-recorded, transcribed, and analyzed by the designer (second author). She collaborated with three participants separately (see Fig. 1) to discuss their interests and personal stories.

In this study we adapted the *designing for one* process to cater to the different experiences of the participants, and allow those to feed the design concept. The designer conducted two *designing for one* processes in parallel (Fig. 1—left side) which converged into a single design proposal. This concept was suitable for both Adele and Bruce. Later on, we continued with this design

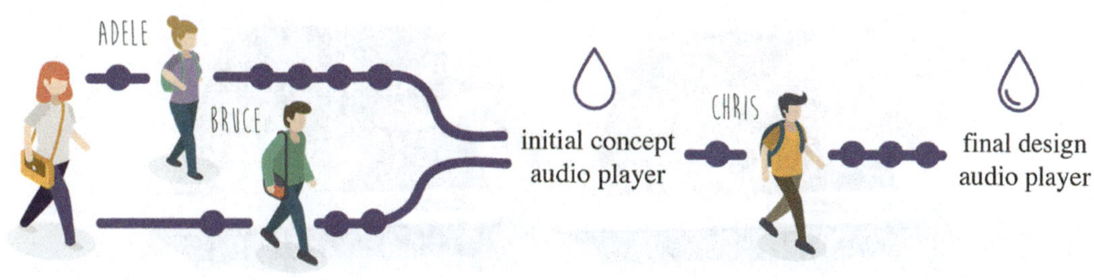

Fig. 1 A visual of the design process, and how the designer interacted with the three participants, the circles represent the number of times the participants were visited

with a new participant, Chris, to explore the personalisation and the design further (Fig. 1—right side).

In phase 1 the process with Adele and Bruce had four sessions: (1) getting to know each other, (2) discussing the idea, (3) testing the interaction of the prototype, and (4) evaluating the audio player in the home context (see Fig. 1). For Adele, one session was added to install the prototype at home evaluation, and Bruce indicated he wanted to stop partaking after the third session, which we will elaborate on in phase 1.

In phase 2 of the project, we aimed to reflect on the design proposal and adapt it with Chris, this phase consisted of four sessions with a similar set-up as described above in phase 1. However, since the prototype advanced more, step 3 and 4 were combined and after going through the interaction of the prototype, the prototype remained at the participants' house for the evaluation.

2.2.1 Final Design: Audio Player 'Your Moments'

In this section we will present our final design 'Your moments', which was the result of the design processes with Adele, Bruce, and Chris. This is anaudio player to stay connected with who you are as a person through memories of valuable moments in your life. The audio is personalized and could for example be poems, music, messages from others, books, or personal memories. Currently, the designer used a voice-over for the personal stories, this worked especially well and deserves further investigation.

Using 'Your moments' starts with turning on the product by touching the button in the middle (Fig. 2—left), then a brief auditory explanation is given on the products' functions and its content. All buttons emit different coloured light, but when the user selects one (Figure—right), this particular one remains coloured while the others turn to white, this indicates which audio file is selected. Then the audio starts playing, and to stop the audio, the user can touch the same button again. Understanding how the final design looks and works supports the result section of phase 1 and 2.

2.3 Results of Phase 1: Two Design for One Processes

Session 1—Getting to Know Each Other. The design researcher introduced herself, the project, and the projects' timeline. Here the session focused on gathering information with regard to understanding the participants, their history, whom they are close with, and what they value. Furthermore, insights were gained about preferred activities using the Leisure Time Canvas [24], which is a storytelling tool to sort activities in terms of preference. Based on the collected information, participant portraits were created.

Session 2—Discussing the Idea. Based on these findings, creative sessions were organized resulting in the idea of memory stones to capture personal memories. A first prototype was created to discuss possible interaction and content with Adele and Bruce separately using keyword cards [25].

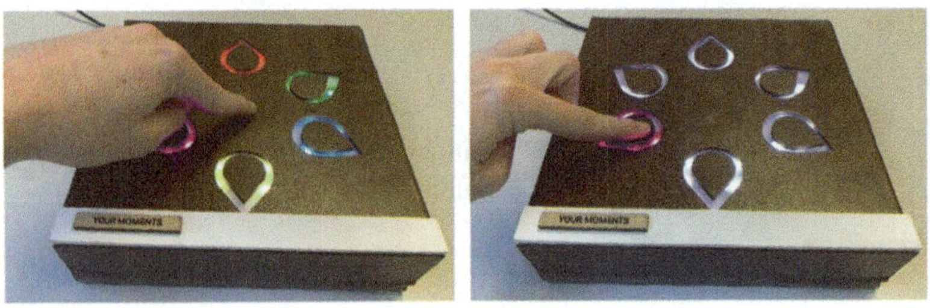

Fig. 2 'Your moments' turns on by touching the middle button (left) and when touching one button, the other buttons turn white (right)

Fig. 3 The prototype evolution of what we brought in the design process for session 2 (left), session 3 (middle), and session 4 (right)

From session 2, we have learned that the product should become more intuitive, the stones were hard to identify and that the tangible elements were appreciated (Fig. 3 left). To process these findings, several changes were made to the prototype for session 3, to easily distinguish the audio files, the stones were colour-coded, placed on a high-contrast surface, and had clarifying texts on the tokens to describe the type of audio (Fig. 3 middle). When asking Bruce about where he would use the design proposal in the future, we noticed that he had a challenging with imaging a potential use case.

Furthermore, we learned that the creating and selecting audio content should be flexible and based on what someone finds most important and reflecting their interests. From the keyword cards, we learned about the audio content, e.g., that Adele ranked fantasy low, while memories, parents, and milestones were high. From Bruce, we learned fantasy and poems were low, while people and experiences were ranked high. At last, Adele and Bruce were given a voice recorder to record a story about a meaningful moment in their life, which we envisioned to be one of the audio files in 'Your moments'.

Session 3—Testing the Interaction of the Prototype. After concretizing our findings from session 2 the designer translated them into a functional prototype. A coloured token, as part of the design in session 2, could be placed on the central platform of the prototype, and by using NFC technology audio files were played on a laptop nearby.

In this session, the Geneva Wheel of Emotions [26] was used to measure emotional reactions to the prototype. For example, Bruce's response on his audio file was positive as it stimulated 'happiness, pride and amusement': '*Listening to the story of the event [a story about going to a fair with his friends] makes me nostalgic, and I find it a funny story to tell, which I do every time in a café as well.*'. However, intertwined with this positive feeling, it reminded him of a recent event that one of his best friends from this story was not feeling well. Also, in the session with Adelle we saw an emotional moment, as she realized her forgetfulness through one audio file. In this audio file she went for dinner with friends, but she did not remember exactly anymore. Reflecting on these findings in the design process, we decided that we did not want the audio player to only be a stringent reminder of loss, and therefore we wanted to explore other audio samples as well. However, we did not want to remove these memories either, as they can also encourage meaningful, emotional, and vivid responses.

Subsequently, by conducting a short usability evaluation [27], we found a challenge in placing a token in the middle of the prototype to play an audio file (see Fig. 3 middle). Besides, in our prototype, to stop playing a message, another token should be placed instead, which was inconvenient.

At last, to close the session, the designer listened with the participants to the recordings. In this activity we found that hearing their own voice was perceived as unpleasant. To overcome this, we used an anonymous voice-over to read out their written stories.

After session 3, Bruce decided he did not want to participate in the final session anymore. He mentioned that he helped the designer already enough, and was happy with his participation and contribution so far. Somehow, he became suspicious to be 'labelled' as a person with dementia due to the research. Since Bruce valued his earlier contribution we kept the data of the earlier sessions in this paper. Bruce's response may indicate a concern of stigmatization so, we, of course, respected his decision to cancel the final session, and will reflect on this in our discussion.

Session 4—Evaluating the Audio Player in the Home Context. With the insights from session 3, we improved the ease of interacting with the design, so the user only needs to touch one of the tokens (button) to start the audio. The prototype was made to work stand-alone and personal audio files were gathered and uploaded.

To study the prototype, we conducted a home evaluation with this prototype (see Fig. 3 right). Adele wrote down her experience of using the prototype at home in a user diary for four days [28]. The diary consisted of four daily questions about her experiences of when she used it, what she listened to, why she chose this audio, and what experience she had while listening.

From this evaluation, we learned that Adele used the audio player when she wanted a moment for herself. She perceived the music of her uncle's band as a beautifully nostalgic reminder to him. Although hearing her voice was sometimes perceived unpleasantly, however, she did perceive the voice-over as pleasant.

2.4 Results of Phase 2: Exploring Personalization with a New Participant by Continuing with the Earlier Results of Phase 1

The design which has been studied at Adele's home was introduced in with Chris, yet a spoken manual was added instead of a written manual. By bringing it to a new participant, we wanted to find out if the design would be meaningful to someone else as well and how it could be personalized further.

We describe here the outcome of session 1 and 4, of which the set-up was similar to Adele's sessions. We had one additional session to prepare for the home study to determining the content of the audio files.

During the first session, we learned that Chris likes to listen to music more (as opposed to memories) which we could facilitate with our design. To determine the content, Chris was asked to share some stories with us. He did this by bringing CD's into the sessions to select songs that he would like to be on the audio player.

For the final session, Chris tested the audio player in his home for five days, and we learned that Chris enjoyed singing along with his favourite music and that the music positively reminded him of previous experiences and joyful memories. The device was also used together with his spouse and the design facilitated a shared experience. In addition, we found that from the audio files accessible, Chris wants to keep access to the memory tokens, while the tokens with music may be changed from time to time.

Debriefing with the Involved Professional Caregiver. After the phases we also debriefed with the professional caregiver. The designer understood from her that all participants proudly shared being a part of the design process with their friends. Furthermore, they really enjoyed working together with the designer. For the prototype, we learned from her that it would be clearer to (1) let the audio file state its content through a spoken message before it starts playing, and (2) make the tokens more distinguishable. These changes were included in a final design proposal.

3 Discussion

In our results we report on the co-design session with three participants, leading to the design proposal 'Your moments'. We indicated how we included the feedback and experiences from the participants in the design. Through this, we aimed to understand how to create a suitable design for the unique needs and lives of the participants. In addition, we learned from our participant Bruce who withdrew from our research and the stakeholders who enable using the design at home.

We aimed to include people with various life experiences in our design process [5]. We did so by taking inspiration from the *designing for one* process [22], and by incorporating the participants' perspectives over several participatory sessions. We learned about their interests, backgrounds, and social environments. In addition we discussed tangible, concrete, and experiential ideas which is important when working with people with memory challenges [12]. This led to uncover various questions with regard to the future development of such audio experiences, such as (1) who is involved in adding the audio and listening to it, (2) what types of audio is preferred in general, and (3) when and how are the audio fragments played. We showed how our design proposal 'Your moments' empowers people by being able to access personal audio and music content. Because the design is open, and serves as a platform to listen to audio files, it requires the users to appropriate it to their preference. People can choose their personal moments with 'Your moments' and have an experience tailored to their interests and identity.

In addition we want to reflect on our experience with participant Bruce as well. We got to know Bruce as a very honest man, and we learned from his directness towards the designer: '*I [Bruce] don't want you to expect miracles from me. I cannot structure my thoughts very well anymore*'. Unfortunately, he did not want to participate in the final session at home, because on the one hand, he found he helped the designer enough and mentioned it was time for him to do something else, and on the other hand. Because he was afraid of the label dementia. We became well-aware that people may not or cannot express their feelings of discomfort that easily always, and thus, keeping track of this stays important in every step of the design process. However, we did learn from the involved professional caregiver that he proudly shared his participation with his friends at the care facility, and also allowed us to use his data in this paper. Yet because he expressed that 'others' would find out about 'his' dementia, we should perhaps have made it even clearer that we would process his results anonymous, and that this research is set up a safe space. Bruce did not mention this, but perhaps this did not come across clear enough, and researchers need to pay attention to this when working with people with memory challenges. Also, in future research we will be more aware of fluctuating moods of participants and make sure that, in the end, they conclude their participation as feeling valued both by the researchers.

The design process of 'Your moments' also contributes to the discussion of how other stakeholders play an important role in supporting people with memory challenges and facilitating a positive experience with technology. In our process for example, during session 3, the participants brought a story in print, which was captured as an audio file via a voice-over. This voice-over of their own story elicited a very positive experience in our participants. This shows how authentic input from the person with memory challenges, became available via the voice-over and 'Your moments' to the person. It could potentially be a novel way to empower people with memory challenges to experience autobiographical content deserving further investigation.

Eventually, we aim to empower users to experience a meaningful moment, which in our case was facilitated by listening to personalised audio content, and it should be explored further what the role of the surrounding social network of users could be with regard to finding, collecting and experiencing audio content. We saw that our third participant Chris included his partner in the experience, and other research shows this as well. For example, in the messaging radio, family and friends can easily send audio recordings using WhatsApp [29]. In this case, the audio content is specifically recorded for the user by his or her social network, which might increase the meaning of the messages. Another interesting example with audio is using both personal and generic content to support personal interests and cater for curiosity [30]. In this case people could first explore personal interests and expand on that with content that is suitable for the general audience.

To conclude, in this study, we have seen that the highly individual focus enabled us to learn in-depth about people's characteristics, which provided creative input to create a design that catered to the personal needs of the participants. In the future, we would like to determine whether this personal approach can be extended to a broader audience if people could personalize a design by themselves and to what extent this approach can inspire researchers to explore personalization in design further.

References

1. Plassman, B.L., Langa, K.M., Fisher, G.G., Heeringo, S.G. Weir, D.R., Ofstedal, M.B., …, Wallace, R.B.: Prevalence of cognitive impairment without dementia in the United States. Ann. Intern. Med. **148**, 427–434 (2008)
2. Buchanan, J.A., Christenson, A., Houlihan, D., Ostrom, C.: The role of behavior analysis in the rehabilitation of persons with dementia. Behav. Ther. **42**, 9–21 (2011)
3. Brankaert, R., den Ouden, E.: The design-driven living lab: a new approach to exploring solutions to complex societal challenges. Technol. Innov. Manag. Rev. **7**, 44–51 (2017)
4. Purves, B.A., Phinney, A., Hulko, W., Puurveen, G., Astell, A.J.: Developing CIRCA-BC and exploring the role of the computer as a third participant in conversation. Am. J. Alzheimers. Dis. Other Demen. **30**, 101–107 (2015)
5. Ludwin, K., Capstick, A.: Using Participatory video to understand diversity among people with dementia in long-term care. J. Psychol. Issues Organ. Cult. 5, 30–38 (2015)
6. Lazar, A., Cornejo, R., Edasis, C., Piper, A.M.: Designing for the third hand : empowering older adults with cognitive impairments through creating and sharing. Proc. Des. Interact. Syst. Conf. (DIS '16), 1047–1058 (2016)
7. Piper, A.M., Cornejo, R., Hurwitz, L., Unumb, C.: Technological caregiving : supporting online activity for adults with cognitive impairments. CHI., 5311–5323 (2016)
8. Brankaert, R., Suijkerbuijk, S.: Outdoor life and technology with dementia. In: Astell, A., Smith, S., Joddrell, P. (eds.) Using technology in dementia care: a guide to technology solutions for everyday living, pp. 53–64. Jessica Kingsley Publishers (2019)
9. Fontana, A., Smith, R.W.: Alzheimer's disease victims: the "unbecoming" of self and the normalization of competence. Sociol. Perspect. **32**, 35–46 (1989)
10. Kanter, C., D'Agostino, N.M., Daniels, M., Stone, A., Edelstein, K.: Together and apart: providing psychosocial support for patients and families living with brain tumors. Support. Care Cancer. **22**, 43–52 (2014)
11. Kitwood, T.: Dementia reconsidered: the person comes first. Buckingham, Open University Press (1997)
12. Cowdell, F.: Preserving personhood in dementia research : a literature review, pp. 85–94 (2006)
13. Houben, M., Brankaert, R., Bakker, S., Kenning, G., Bongers, I., Eggen, B.: Foregrounding everyday sounds in dementia, pp. 71–83 (2019)
14. Treadaway, C.: Personalization and compassionate design. In: Brankaert, R. and Kenning, G. (eds.) HCI and Design in the Context of Dementia, pp. 49–61. Springer (2020)
15. Baird, A., Thompson, W.F.: The impact of music on the self in dementia. J. Alzheimer's Dis. **61**, 827–841 (2018)
16. Shinohara, K., Wobbrock, J.O.: In the shadow of misperception: assistive technology use and social interactions. Conf. Hum. Factors Comput. Syst.—Proc., 705–714 (2011)
17. Mattelmäki, T., Battarbee, K.: Empathy probes. In: Participatory Design Conference, pp. 266–271 (2002)
18. Wallace, J., McCarthy, J., Wright, P.C., Olivier, P.: Making design probes work. Proc. SIGCHI Conf. Hum. Factors Comput. Syst.—CHI '13, 3441–3450 (2013)
19. Lindsay, S., Jackson, D., Ladha, C., Ladha, K., Brittain, K., Olivier, P.: Empathy, participatory design and people with dementia. In: CHI'12, pp. 521–530 (2012)
20. Astell, A.J., Malone, B., Williams, G., Hwang, F., Ellis, M.P.: Leveraging everyday technology for people living with dementia: a case study. J. Assist. Technol. **8**, 164–176 (2014)
21. Sanders, Stappers, P.J.: Co-creation and the new landscapes of design. CoDesign. **4**, 5–18 (2008)
22. Wilkinson, A., Hendriks, N., Stones, C.: Designing for one; how adjusting variables influenced design student creativity. In: ICDC (2018)
23. Haan, M.C. den, Brankaert, R., Lu, Y.: Design for one: personalisation and experiences of design researchers and participants. In: ICDC (2020)
24. den Haan, M., Brankaert, R., Lu, Y.: The leisure time canvas: eliciting empathy for older adults through activities and hobbies. In: Academy for Design Innovation Management (2019)
25. Thomas, B.J., Sekar, P.C.: Measurement and validity of jennifer aaker's brand personality scale for colgate brand. Vikalpa. **33**, 49–61 (2008)

26. Scherer, K.R.: What are emotions? And how can they be measured? Soc. Sci. Inf. **44**, 695–729 (2005)
27. Brooke, J.: SUS: A quick and dirty usability scale. In: Jordan, P.W., Thomas, B., McClelland, I.L., Weerdmeester, B. (eds.) Usability Evaluation in Industry, pp. 189–194. London, Taylor and Francis (1996)
28. Bartlett, R.: Modifying the diary interview method to research the lives of people with dementia. Qual. Health Res. **22**, 1717–1726 (2012)
29. Wintermans, M., Brankaert, R., Lu, Y.: Together we do not forget: co-designing with people living with dementia towards a design for social inclusion. Des. Manag. Acad. (2017)
30. Thoolen, M., Brankaert, R. G. A., Lu, Y.: AmbientEcho: exploring interactive media experiences in the context of residential dementia care. In: AmbientEcho: Exploring Interactive Media Experiences in the Context of Residential Dementia Care (blz. 1495–1508). Association for Computing Machinery, Inc. (2020). https://doi.org/10.1145/3357236.3395432

Lessons Learned from Remote User-Centered Design with People with Dementia

Emma Dixon, Ashrith Shetty, Simone Pimento, and Amanda Lazar

Abstract

Modified approaches to user-centered design have been developed to better accommodate the unique needs of people with dementia within the design process. These approaches were developed to address challenges with abstract thought and verbal communication of people with dementia, as well as bridge the gap in the everyday life experiences of designers to those of people with dementia. With these new methods to better facilitate design work with people with dementia, we are missing an understand of how to conduct such design work remotely. In this paper we present a case-study of the use of remote methods for a user-centered design process which included interviews, low-fidelity prototyping sessions, and asynchronous evaluation of prototypes with tech-savvy people with mild to moderate dementia. The main contribution includes a discussion of the benefits and challenges of using this remote method with tech-savvy people with mild to moderate dementia.

Keywords

User-centered design · Remote methods · Dementia

1 Introduction and Background

A central tenet of many Human-Computer Interaction methods is ensuring the population for which a technology is designed is involved in the design process. In working with people with dementia, this is especially key, for reasons ranging from the reliance on caregivers as proxies [1] to the challenges in supporting technology adoption and a sense of empowerment [2].

Yet, user-centered design approaches often require modification when being applied with populations with dementia. Researchers have devised new approaches to overcome some of the barriers to designing in dementia, such as the wide gap in the everyday life experiences of designers to those with people with dementia. The KITE approach has been designed to foster empathy in relationships between designers and people with dementia [3]. Similarly, the OASIS method was used to establish "an appropriate atmosphere in the meetings" [3] to bridge the gap between younger designers' technological insight and need areas for future technology in older participants' daily lives [4]. The OASIS method

E. Dixon (✉) · A. Shetty · S. Pimento · A. Lazar
University of Maryland, College Park, MD, USA
e-mail: eedixon@umd.edu

R. Brankaert et al. (eds.), *Dementia Lab 2021: Supporting Ability Through Design*, Design For Inclusion 2, https://doi.org/10.1007/978-3-030-70293-9_6

consists of four stages: (1) stakeholder identification and recruitment (2) presenting participants with a video illustrating a scenario with future intervention for the problem domain (3) exploratory meetings which define the domain of the project and user requirements, and (4) low-fidelity prototyping sessions [3, 4]. In the work we present in this paper, we took a similar approach of integrating prototyping with video presentations—but in an entirely remote approach.

Past work has found that some of the conventional approaches to brainstorming and generating design ideas can be difficult for people with dementia. Bourazeri attempted to use fill in the blank cards to spur inspiration for the design of sensor technology in participant's homes [5]. However, this method proved unsuccessful as participants had difficulty understanding how the sensors would work and were unable to think abstractly on the subject [5]. This finding aligns with Hendrik et al.'s work which found that participants with dementia had a hard time making abstractions from visuals of how technology would work [6]. Though this work points to the deficiencies of a visual-based approach in co-design with people with dementia, other researchers have found people with dementia are able to think abstractly using symbols and metaphors during co-design [7]. Although not tested with people with dementia, one method to overcome the potential difficulty of abstraction is to use high-fidelity prototypes which reduce the amount of abstraction needed, which was shown to be useful for users with aphasia—a condition which affects people's ability to communicate verbally [8–10].

In addition to challenges with abstract thought in co-design with people with dementia, challenges with verbal communication has been a subject of interest for research that questions the usefulness of traditional approaches to co-design sessions with this population [11–13]. Foley et al. suggest using more embodied interactions in the design process to overcome these verbal speaking challenges [12, 14]. Neate et al. found that using the personas allowed people with aphasia to "optimize their use of language rather than having to use extensive language to articulate a particular thought" [9]. In contrast, Bourazeri et al. had their co-designers with dementia describe their design ideas out loud while the researcher would draw out the ideas and the group would actively critique the design being drawn [5]. To our knowledge, this is one of the only examples of how low-fidelity prototyping in co-design has been adapted for users with dementia [5].

In the study we report on here, we leveraged past findings about engaging in user-centered design with people with dementia to test a novel remote approach. The need for remote approaches is especially applicable for people with dementia as many individuals living with the condition are over the age of 65 and therefore at a much higher risk of suffering severe illness from COVID-19, making in-person research unsafe for the foreseeable future. Additionally, remote methods can be beneficial as a way to allow participants the ability to complete studies in their natural work or home environment rather than an academic or laboratory environment [15, 16]—this is particularly useful in dementia given the importance of contextual cues and routines [17, 18]. When it comes to technology design, researchers do at times include people with dementia in remote interviews and field trials (e.g., [19, 20]). However, co-design, which often centers around interactions with design materials and prototypes, is much more challenging to translate online for dementia.

This paper presents a case-study of the use of remote methods for a co-design process which included interviews, low-fidelity prototyping sessions, and asynchronous evaluation of prototypes with tech-savvy individuals with dementia. Our work focuses on people with mild to moderate dementia, the stage of the condition where individuals experience difficulty with remembering names, word finding, performing tasks, and planning [21] but with the right support are generally still able to manage activities of daily living. The main contribution of this paper includes a discussion of the benefits and challenges of using this method with tech-savvy people with mild to moderate dementia.

2 Method

Below, we describe our approach to remote data collection and analysis. We also provide information on participant recruitment and demographics.

2.1 Data Collection and Analysis

All procedures were approved by the university's Institutional Review Board. For this study we took a general User-Centered Design approach [22]. See Fig. 1 for a visual of the process we followed and who was involved in each step.

We first followed the approach of learning more about participants' lives and needs through "empathizing interviews"—remote semi-structured interviews to understand current technology usage and any issues participants experience. Our goals for conducting these interviews was to scope the scenarios which we would design for in the co-design sessions as well as get participants' ideas for future technologies. Each interview and observation session was between thirty minutes and an hour in length. The semi-structured format [23] allowed us to guide the participants into relevant discussions while making sure we were not controlling the narratives of conversations. We asked probing, open-ended questions to uncover deeper insights into their experiences. All four interviews were conducted remotely and were audio/video recorded. Observation notes were taken throughout each interview. The audio recordings were later transcribed

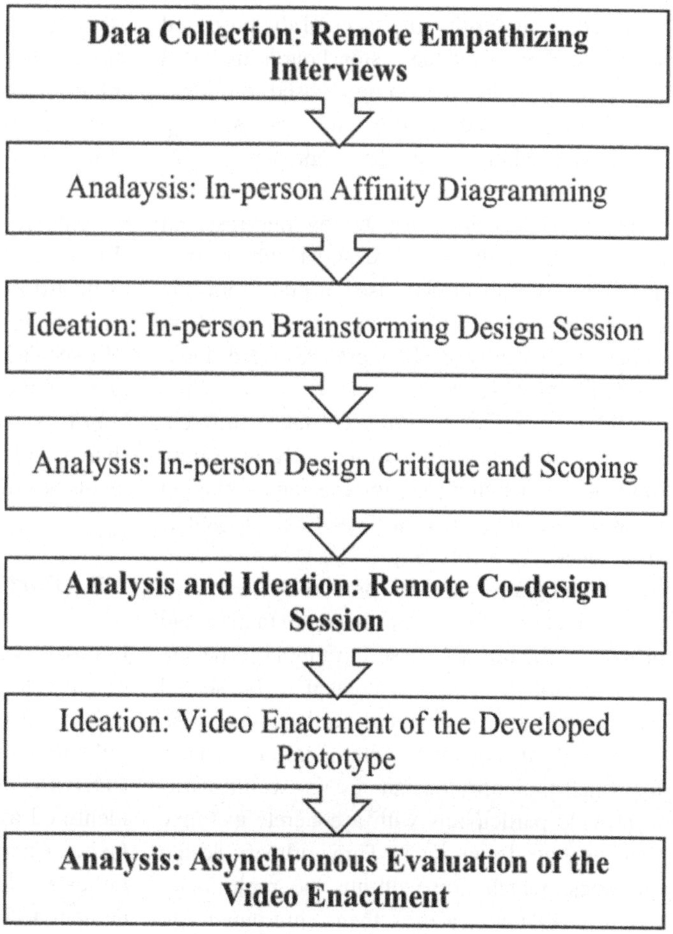

Fig. 1. Study process. phases that are bolded are those that people with dementia are included

for analysis using Otter.ai. Participants were compensated with a $10 Amazon gift card.

The research team then used affinity mapping to identify the emergent themes among the findings and insights we gained in the interviews [24]. The research team conducted interpretation sessions on each interview, taking notes from the transcribed interviews and translating each insight to individual sticky-notes. The team then grouped the notes, to reveal emerging themes most relevant to the design problem [24]. The affinity map was created on Miro where we grouped the participants' needs into three major focus areas: social needs, sensory needs, and technology needs. The research team gathered for a preliminary brainstorming design session where we discussed design ideas for each of these major need areas. We created storyboards that illustrated commonly occurring problem scenarios in the lives of people with dementia, as well as designs for the interfaces of the proposed solutions. Several of these storyboards included sketches of future technology based on ideas participants shared during the empathizing interviews, taking a similar approach to co-design as in past work (e.g., [5]). These designs were then internally critiqued by our research team, where only those designs which best addressed the identified user needs from the empathizing interview and were most novel (in terms of gaps in research literature) were documented in Google Docs.

Different research projects take different approaches to co-design, from end users doing most of the sketching [3] to researchers playing more of the design role with participants generating ideas and feedback [5]. We took the latter, by sketching design ideas previously described by participants. As a way to contextualize and provide structure to these preliminary design ideas, a Google Doc was created to be shared with participants during the co-design session. We started this document with an initial scenario for which the entire design session would focus to provide participants with a concrete example to focus their design ideas. Three primary design questions, which arose during the preliminary design session, were then presented to participants to provide structure for the co-design sessions. For each of these three design questions we presented participants with two example designs, in the form of sketches and story-boards, for their critique and to iterate on. The document was formatted to include one design per page so as to avoid visual overstimulation.

Two members of the research team then conducted a remote co-design session with one participant with dementia via Zoom video-conferencing based on these preliminary designs. This session was approximately one hour long and was audio/video recorded. Before the session, the participant was asked to have a pen and paper handy and was encouraged to sketch and explore their own ideas during the session. During the video call, we shared our screen with the participant (see Fig. 2) to walk them through the various scenarios using our storyboards. The participant was encouraged to ask questions, critique the solutions, and also suggest their own ideas and insights into how the solutions could be further improved. The participant was compensated with a $20 Amazon gift card.

Incorporating the feedback gained in the remote co-design session, we developed a Wizard of Oz prototype of our design solution using Adobe XD. We then used the prototype to create a video-enactment to showcase the problem scenario and to show how the various features of the proposed prototype solution would help to solve the problem. This video was then shared with two of the original interview participants for discussion and evaluation.

2.2 Participants

In order to qualify for the study, participants had to have some form of dementia and use technology on a regular basis. Four individuals with mild to moderate dementia participated in this study (age range 57–67, average age 61.5; three identified as female and one as male; all identified as Caucasian). The age range of participants indicates that they likely have younger onset dementia [25]. Table 1 has additional details on

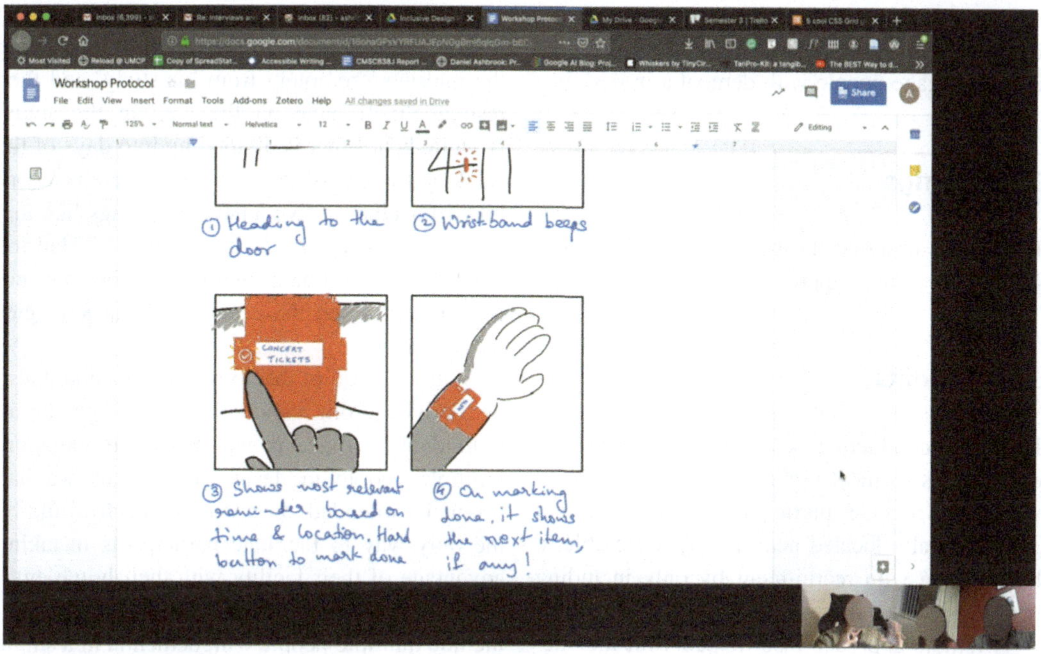

Fig. 2. Google document of remote co-design session

Table 1. Demographic information

Participant pseudonym	Age	Gender	Type of dementia	Stage of dementia	Study phase participated in
Jessica	59	Female	Vascular dementia/White matter disease	Mild/Moderate	Interview
Amelia	67	Female	Major neuro-cognitive impairment	Mild/Moderate	Interview
Rose	57	Female	Younger onset alzheimer's	Mild	Interview, video evaluation
Joe	63	Male	Mixed vascular dementia/alzheimer's	Mild/Moderate	Interview, co-design, video evaluation

demographics, including: age, gender, type of dementia, stage of dementia (which was self-reported), and the specific phases of the study each participant participated in. Jessica, Amelia, and Rose all resided in the US and Joe in the UK.

In regards to the phase of the study each participant participated in, two of the four initial participants were recruited to participate in remote co-design sessions, Rose and Joe. These participants were chosen after scoping the project to location based reminder systems because of the relevance of this scope to their current reminder systems. Unfortunately, due to medical complications, Rose was not able to complete the co-design session. Therefore, only Joe completed the remote co-design session, but Rose and Joe both completed the evaluation phase of the video-enactment of the wizard-of-oz prototype.

Given the small number of participants and the specifics of their demographics, the goal of this paper is not to advance our understanding of reminder systems for dementia: rather, it is to understand some of the benefits and challenges that arise when working towards remote approaches to engage people with dementia in design.

3 Findings

Below, we describe the benefits and challenges in taking a remote approach to design.

3.1 Benefits

Remotely conducting sessions was helpful for enabling us to meet our recruitment criteria, as we could include participants who were not geographically located near us. We were able to be selective with recruitment by only including participants who were avid technology users.

A remote approach also helped with the successful facilitation of the study. Participants were able to talk to us from a setting in which they were comfortable with the environment. Additionally, by conducting the study with participants in their own homes, using their own technology, there were no barriers to using the digital tools that we selected for collaboration. In other words, they could access Zoom and Google Docs on their preferred devices rather than needing to use a device that we supply and might be less familiar. For example, Amelia strictly uses Apple products due to the interconnectivity between devices (e.g. iPhone, iPad, Mac laptops) and the "unlimited support" she receives from "go[ing] to the Mac store down the street." If we had asked her to use Windows products for a co-design session this could have been a barrier for participation as she is not used to this system.

A remote approach that allowed individuals to be in their own homes was most essential when it came to asking individuals to talk about their current technology workflows. Because they were situated in their home environment, participants did not have to rely on their memory to recall the various technical and non-technical aspects of their scheduling systems, but instead could just look around, recognize, and discuss them with us. For example, when a researcher asked Joe what other systems he used, he first looked around the room, as he was completing the interview remotely from his study, and then responded: "I have on the wall in the study. A white board that is divided up into days of the week and weeks of the month and there is a cork board for pinning live letters and things like that things that are waiting to be concluded." Had Joe not been in his home environment he may not have remembered these non-technical strategies for reminders.

While several of the above benefits could also be derived from conducting co-design in an individuals' home, if we had been engaging multiple people in design at once as we had originally intended, a remote approach would be the only way to facilitate participants in taking advantage of their facility with their home environments. Further, our remote approach is able to include multiple people with dementia in a single co-design workshop with their technical accessibility configurations in their home environment, such as Joe's use of a mounted tv attached to his computer to magnify his screen to his needed level. This level of accessibility configuration for several people in a group in-person co-design session would be costly and difficult for researchers to recreate in a lab setting.

3.2 Challenges

Although there were positive aspects of remote interviews and observation sessions, which went smoothly overall, there were several downsides of this method. While participants were able to access their own devices and setups in their homes, it was difficult for us to observe some kinds of technology use. In particular, it was difficult to see how people used their phone applications for scheduling because they could not use the applications and simultaneously show us their screen of what they were doing with that application. They had to keep flipping the phone towards the computer camera to show us their use, which was not ideal.

A second challenge arose when, in sharing their screen to demonstrate their scheduling systems, participants unknowingly shared personal information of both their own and their friends. For instance, Joe keeps a record of interactions he has with other people, details about their lives and families, as well as when he was going to meet them again, all in a spreadsheet database to act as a memory aid. Although this was a very interesting and thought provoking example of a personal scheduling system, we were unable to keep or analyze the video-recording of this interaction to ensure the privacy and anonymity of this data as those other people had not consented for this information to be shared.

A third challenge arose in attempting to engage Joe in the co-design session to comment on our preliminary design ideas. We opened the discussion asking him to feel free to ideate and sketch out his own designs/solutions. Although we were able to capture Joe's thoughts on each design, he did not sketch anything of his own. When we asked why he didn't sketch his ideas out, Joe explained "if we'd hit a point where I was trying to put something across and my words were capable of giving you a clear enough picture of what I was seeing in my head, then yeah, maybe being able to draw." He did however still engage in co-design through critiquing certain aspects of the preliminary designs, as well as elaborating on and suggesting new features for our original designs. At the end of the co-design session we asked Joe to critique the remote co-design method. He described the remote co-design session as: "wonderful" and increasing his "motivation, and inspiration" [Joe]. The largely verbal remote co-design session worked for him because "I'm comfortable with technology. And I can, therefore visualize what it is you're saying and showing and that it's comfortable for me." Though we appreciate that Joe found verbal approaches sufficient, we recognize that this is not an approach that will work for many, and we are iterating on our approach to engage people with dementia more actively in the design process.

A final challenge arose when we had participants evaluate the video walkthrough of our Wizard of Oz prototype which incorporated the critiques and suggestions from Joe during the co-design session. Using the prototype, we produced a video consisting of an enactment of the problem scenario as well as how the proposed solution could mitigate the problem. A video enactment of the prototype was used rather than a click-through prototype due to scheduling conflicts with both Rose and Joe, making asynchronous evaluations the most viable alternative. This video was seen as overly complicated by both participants, "even for me that knows what it is about" [Joe]. In the future, we will follow Joe's suggestion to "Pick one specific example of a daily use" and then "Show the example journey" from the need for the system, to set up, to one example in operation [Joe].

4 Discussion

In this paper, we present a case study of engaging people with mild to moderate dementia in a remote co-design process. It is important to note that the relative smoothness of our process may have stemmed from our inclusion criteria, which restricted the population to people with dementia who were tech-savvy. This likely made it easier for them to use the videoconferencing system that we selected, but also may have made it more feasible to discuss abstract design ideas (in contrast to past work [5, 6]). Additionally, it's important to note our participants were in the mild to moderate stages of dementia where they were all able to communicate verbally to express and critique design ideas. Even with the consideration for stage and technological literacy, our findings demonstrate that it is entirely feasible for tech-savvy people with mild to moderate dementia to translate visual representations to technology concepts and critique/improve upon these concepts to tailor them for their own use.

Though past work stresses the importance of providing embodied and sensory-rich ways for people with dementia to engage in design, two instances in our data seem to contrast with these past findings. First, the participant who was a part of the interactive design session preferred to relay their ideas verbally, rather than use the approach we offered to draw. This may be due to the stages of people with dementia participating in this work, as embodied interactions may be more feasible as people advance in dementia. Further research is needed to uncover methods that may better invite people at all stages of dementia to remotely engage in hands-on activity during the design process (for one approach, see [26]). A second issue arose when we attempted to provide a sensory rich depiction of how the prototype system might work, with a video rather than a verbal description. The video was seen as overly complicated and confusing, demonstrating the limitations of video prototypes. We offer considerations to improve this piece below.

The design process involved both synchronous portions and an asynchronous component, when individuals viewed the video that we created. Interestingly, the most negative and critical feedback we received on the proposed solution and the process came during the asynchronous video evaluation. We believe this could be due to the fact that participants were more comfortable sharing their criticisms because they didn't need to say them directly to the designer. This provides one benefit to asynchronous interaction, which we had previously tried to avoid due to the importance of real time interaction in dementia.

However, additional iterations of this video-based evaluation stage are necessary, as the majority of the criticisms were on the accessibility and understandability of the video itself, rather than the prototype. The video was 4 min long, was very detailed, and was narrated by a non-native English speaker. It is possible that the speed of the video or the lack of subtitles made the video inaccessible and difficult to understand. Potentially this could have been due to the multi-sensory format of the video causing sensory overload [27] as the video included audio descriptions paired with text to be read on the screen of what was being input into the system, as well as visual displays of the system. Additionally, participants expressed they wanted a quick overview of the product rather than us show them every function we designed. However, this may not provide the detailed evaluation that designers need to be sure the product is usable by users with dementia. For this reason, having participants themselves use the click through prototype or conduct a cognitive walk-through of the high-fidelity prototype themselves rather than watching a video tutorial may be better for evaluation of each of the functions of a prototype—as shown in prior research with users with Aphasia [8–10]. In a remote session this could be done by participants taking remote control of the researchers computer to evaluate a click-through prototype.

In this paper we have presented a case-study of the use of remote methods for a co-design process, including interviews, a low-fidelity prototyping session, and asynchronous video evaluation of the resulting prototype with tech-savvy individuals with mild to moderate dementia. The main contribution of this paper includes a discussion of the benefits and challenges of using this remote method. Our work indicates that remote methods may be a feasible approach to designing with some people with mild to moderate dementia, though more work is needed to understand how to remotely test working prototypes, facilitate hands-on activities during the design process, and broaden participation beyond tech-savvy individuals with dementia. With consideration for the next generation of people with dementia who will be familiar with remote working tools, our findings provide a glimpse into the future feasibility of remote co-design methods with people with mild to moderate dementia.

References

1. Span, M., Hettinga, M., Vernooij-Dassen, M., Eefsting, J., Smits, C.: Involving people with dementia in the development of supportive IT applications: a systematic review. Ageing Res. Rev. **12**(2), 535–551 (2013). https://doi.org/10.1016/j.arr.2013.01.002
2. Greenhalgh, T., Jackson, C., Shaw, S., Janamian, T.: Achieving research impact through co-creation in community-based health services: literature review and case study. Milbank Q. **94**(2), 392–429 (2016). https://doi.org/10.1111/1468-0009.12197
3. Lindsay, S., Brittain, K., Jackson, D., Ladha, C., Ladha, K., Olivier, P.: Empathy, participatory design

and people with dementia. In: Proceedings of the 2012 ACM annual conference on Human Factors in Computing Systems—CHI '12, 521 (2012). https://doi.org/10.1145/2207676.2207749
4. McNaney, R., Balaam, M., Holden, A., Schofield, G., Jackson, D., Webster, M., Galna, B., Barry, G., Rochester, L., Olivier, P.: Designing for and with people with parkinson's: a focus on exergaming. In: Proceedings of the 33rd Annual ACM Conference on Human Factors in Computing Systems (CHI '15), 501–510 (2015). https://doi.org/10.1145/2702123.2702310
5. Bourazeri, A., Stumpf, S.: Co-designing smart home technology with people with dementia or Parkinson's disease. In: Proceedings of the 10th Nordic Conference on Human-Computer Interaction—NordiCHI '18, 609–621 (2018). https://doi.org/10.1145/3240167.3240197
6. Hendriks, N., Huybrechts, L., Wilkinson, A., Slegers, K.: Challenges in doing participatory design with people with dementia. In: Proceedings of the 13th Participatory Design Conference on Short Papers, Industry Cases, Workshop Descriptions, Doctoral Consortium papers, and Keynote abstracts—PDC '14—volume 2, 33–36 (2014). https://doi.org/10.1145/2662155.2662196
7. Price, D., Jacobs, R., Darzentas, D., Vallejos, E.P., Chadborn, N., Martindale, S., Urquhart, L.: MeMa: designing the memory machine. In: Companion Publication of the 2019 on Designing Interactive Systems Conference 2019 Companion—DIS '19 Companion, 271–276 (2019). https://doi.org/10.1145/3301019.3323882
8. Galliers, J., Wilson, S., Marshall, J., Talbot, R., Devane, N., Booth, T., Woolf, C., Greenwood, H.: Experiencing EVA park, a multi-user virtual world for people with aphasia. ACM Trans. Access. Comput. **10**, 4: 15:1–15:24 (2017). https://doi.org/10.1145/3134227
9. Neate, T., Bourazeri, A., Roper, A., Stumpf, S., Wilson, S.: Co-created personas: engaging and empowering users with diverse needs within the design process. In: Proceedings of the 2019 CHI Conference on Human Factors in Computing Systems—CHI '19, 1–12 (2019). https://doi.org/10.1145/3290605.3300880
10. Roper, A., Marshall, J., Wilson, S.: Benefits and limitations of computer gesture therapy for the rehabilitation of severe aphasia. Frontiers in Human Neuroscience 10 (2016). https://doi.org/10.3389/fnhum.2016.00595
11. Brereton, M., Sitbon, L., Abdullah, M.H.L., Vanderberg, M., Koplick, S.: Design after design to bridge between people living with cognitive or sensory impairments, their friends and proxies. CoDesign **11** (1), 4–20 (2015). https://doi.org/10.1080/15710882.2015.1009471
12. Foley, S., McCarthy, J., Pantidi, N.: The struggle for recognition in advanced dementia: implications for experience-centered design. ACM Trans. Comput.-Hum. Interact. **26**(6): 40:1–40:29 (2019). https://doi.org/10.1145/3359594
13. Hendriks, N., Slegers, K., Duysburgh, P.: Codesign with people living with cognitive or sensory impairments: a case for method stories and uniqueness. CoDesign **11**(1), 70–82 (2015). https://doi.org/10.1080/15710882.2015.1020316
14. Kontos, P., Martin, W.: Embodiment and dementia: exploring critical narratives of selfhood, surveillance, and dementia care. Dementia **12**(3), 288–302 (2013). https://doi.org/10.1177/1471301213479787
15. Rex Hartson, H., Castillo, J.C., Kelso, J., Neale, W. C.: Remote evaluation: the network as an extension of the usability laboratory. In: Proceedings of the SIGCHI Conference on Human Factors in Computing Systems (CHI '96), 228–235 (1996). https://doi.org/10.1145/238386.238511
16. Petrie, H., Hamilton, F., King, N., Pavan, P.: Remote usability evaluations with disabled people. In: Proceedings of the SIGCHI Conference on Human Factors in Computing Systems (CHI '06), 1133–1141 (2006). https://doi.org/10.1145/1124772.1124942
17. Lazar, A., Edasis, C., Piper, A.M.: A critical lens on dementia and design in HCI. In: Proceedings of the 2017 CHI Conference on Human Factors in Computing Systems—CHI '17, 2175–2188 (2017). https://doi.org/10.1145/3025453.3025522
18. Morrissey, K., McCarthy, J., Pantidi, N.: The value of experience-centred design approaches in dementia research contexts. In Proceedings of the 2017 CHI Conference on Human Factors in Computing Systems—CHI '17, 1326–1338 (2017). https://doi.org/10.1145/3025453.3025527
19. Kuwahara, N., Abe, S., Yasuda, K., Kuwabara, K.: Networked reminiscence therapy for individuals with dementia by using photo and video sharing. In: Proceedings of the 8th international ACM SIGACCESS conference on Computers and accessibility—Assets '06, 125 (2006). https://doi.org/10.1145/1168987.1169010
20. Lazar, A., Dixon, E.E.: Safe enough to share: setting the dementia agenda online. Proceedings of the ACM on Human-Computer Interaction 3, CSCW: 1–23 (2019). https://doi.org/10.1145/3359187
21. Alzheimer's Association.: Stages of alzheimer's. Alzheimer's disease and dementia. Retrieved October 16, 2020 from https://alz.org/alzheimers-dementia/stages
22. User-Centered Design Basics: Usability.gov. Retrieved February 22, 2020 from /what-and-why/user-centered-design.html
23. Adams, W.C.: Conducting semi-structured interviews. In: Handbook of Practical Program Evaluation.

John Wiley & Sons, Ltd, 492–505 (2015). https://doi.org/10.1002/9781119171386.ch19
24. Holtzblatt, K., Beyer, H.: Contextual design. 94
25. Alzheimer's Association.: Younger/early-onset alzheimer's. Alzheimer's disease and dementia. Retrieved March 16, 2020 from https://alz.org/alzheimers-dementia/what-is-alzheimers/younger-early-onset
26. Maddali, H.T., Dixon, E., Pradhan, A., Lazar, A.: Supporting remote participation when designing with people with dementia. In: Extended Abstracts of the 2020 ACM Conference on Computer-Supported Cooperative Work and Social Computing—CSCW'20 CHI '19
27. Dixon, E., Lazar, A.: The role of sensory changes in everyday technology use by people with mild to moderate dementia. In: Conditionally Accepted in Proceedings of the 2020 ACM SIGACCESS Conference on Computers and Accessibility ASSETS'20 (2020)

Smile: Capturing and Sharing Personal Photos to Stimulate Social Relations and Support Self-Identity in Dementia

Maarten Houben, Veerle van Engen, Gail Kenning, and Rens Brankaert

Abstract

As dementia progresses, it often prompts changes to living arrangements and care provision as people with dementia are required to move from their home environment into long-term care. Over time, family members may lose touch with the everyday lives of people with dementia and no longer share the everyday moments that preserve identity and social connection. In this paper, we present Smile: a camera-based system that captures and shares photos of residents with their caregivers or relatives visiting the dementia care environment. We conducted a four-week in-context study with Smile to explore how this design intervention could support existing relations with relatives and initiate new social interactions with the care staff. Our results indicate that capturing and sharing photos (1) provided the family with insights into the everyday lives of the residents, (2) enhanced the quality of social encounters, and (3) retained dignity and offered a sense of normalcy to the residents. We discuss the implications of using camera-based systems in dementia care to strengthen social relations and provide opportunities to reinforce identity and provide agency through choices made about appearances.

Keywords

Dementia · Design · Person-centered care · Photos · Long-term care

M. Houben (✉) · R. Brankaert
Department of Industrial Design, Eindhoven University of Technology, Eindhoven, The Netherlands
e-mail: m.houben1@tue.nl

V. van Engen
Erasmus School of Health Policy & Management, Rotterdam, The Netherlands

G. Kenning
Ageing Futures Institute, fEEL (Felt Experience and Empathy Lab), University of New South Wales, Sydney, NSW, Australia

R. Brankaert
School for Allied Health Professions, Fontys University of Applied Sciences, Eindhoven, The Netherlands

1 Introduction

The care demands and needs of people with dementia change over time as their condition progresses or they move into long-term care. This transition from a familiar home environment to higher levels of care can be stressful and is linked to decreased social interactions [1]. Informal caregivers and relatives are often conflicted with feelings such as guilt about the transition of their loved one into professionalized

R. Brankaert et al. (eds.), *Dementia Lab 2021: Supporting Ability Through Design*, Design For Inclusion 2, https://doi.org/10.1007/978-3-030-70293-9_7

care environments [2]. While in the home, families have in-built routines to regularly check on their loved ones [3], but these routines cannot be maintained when the person with dementia is transferred to a formal care setting [1]. As a result, family members and relatives lack insight into day-to-day activities and care practices and so can disengage from the care context [4], making it challenging to socially connect with residents during family visits [5].

In response to the decrease in social encounters, professional caregivers aim to engage the residents in the care environment in social interactions during everyday care practice [6]. However, the care staff is challenged to connect socially and emotionally to people living with dementia due to a heavy workload and time constraints in providing practical care tasks [7]. Furthermore, residents with dementia are often unable to share details about their past because of cognitive limitations, and so professional caregivers find it hard to engage meaningfully and at a personal level [6]. This lack of personal connection correlates with higher incidences of behaviors relating to unmet needs or the impact of neuropsychological symptoms of dementia and can negatively affect their social environment and evoke distress for all stakeholders involved [8].

Sharing socially engaging activities, events, or experiences can lead to moments of high-quality connections between people [9]. This feeling of connection promotes positive emotions, can contribute to the quality of life, and can build resilience for both the family member and the person living with dementia [6]. The loss of such feelings of connection and understanding can result in a faster progression of dementia [10]. Interactive technologies have been shown to support meaningful engagement, interaction, and connection for people with dementia, their family members, and care staff [11, 12]. Therefore, this project explores how technology can focus on the person with dementia and support their social engagements with visitors and professional caregivers.

We will discuss the design and evaluation of Smile, a camera-based system for capturing and sharing personal photos (see Fig. 1) in long-term dementia care environments to stimulate engagement and connection between people with dementia, caregivers, and relatives. A person living with dementia can take a photo with a visitor or caregiver using the Smile camera, situated in a communal living area of the care facility. The photo is then automatically sent to a digital photo frame inside the private room of the person living with dementia. The photo supports the person with dementia in recognizing and remembering people who visited them and can provide virtual information for visitors and professional caregivers about what activities occurred throughout the day, providing cues for a personal conversation.

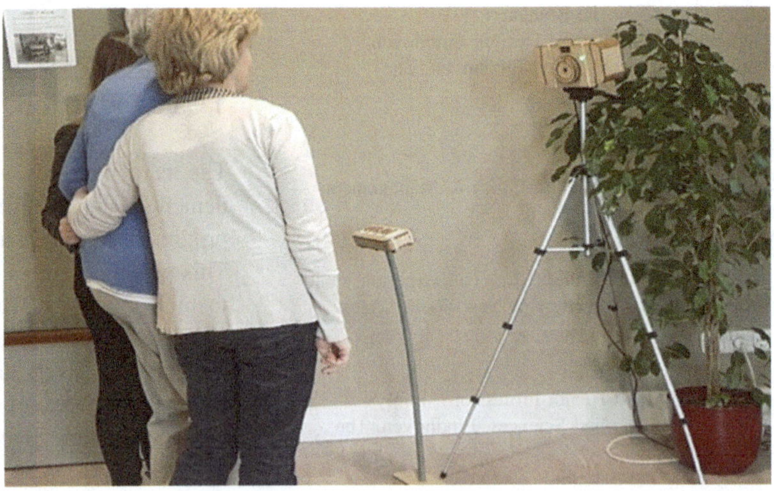

Fig. 1 The smile system provides a camera set up in the communal living area to capture meaningful social moments

We conducted a four-week in-context study of the Smile camera system with four residents with severe dementia and gathered qualitative data through observation and interviews. The results revealed that taking and sharing photos with Smile offered (1) insights into the everyday life of the resident, (2) an increase in the quality of social encounters, and (3) a sense of dignity and normalcy to the lives of the residents. The implications of these findings inform future design research in using camera-based technology and personal photos in dementia care to increase personal connection and social cohesion, and to support identity and self-recognition in people with dementia.

2 Photos in Care Homes

Photos evoke emotional responses in sharing meaningful stories or narratives within groups of friends or domestic households [13]. Sharing personal photos has fostered reciprocal communication and relationships between care managers and their aged clients living independently in their own homes [14]. Research has demonstrated the value of capturing and sharing personal photos for social inclusion, to support care, and to connect with the world outside the care home for older adults in care environments [15, 16]. 'Photostroller' successfully engages older adults in a playful experience by providing categories of generic content on an interactive stroller [15]. Physical photos have been shown to foster communication between family, staff, and older adults in an assisted living environment with low-threshold interaction using a wall projection [16].

For people with dementia, photos have been widely used during reminiscence therapy to prompt autobiographical memories to evoke positive emotional responses and support well-being [17]. Related research in the field of human-computer interaction (HCI) shows that digital social sharing can support people living with dementia [18], and access to personal and generic media in long-term dementia care can facilitate shared experiences and contribute to the personal dignity of residents [19]. In another example, lifelogging cameras or wearable cameras that take photos from a first-person perspective throughout the day continuously can support selfhood and identity by eliciting memories and stimulating reflection about everyday experiences [20]. While the person with memory loss may not recall wearing the camera, discussion about the origins of the photos is richer because of the availability of the photos [17]. Thus, we see that capturing and sharing photos, particularly of care home residents with dementia, can reinforce a sense of self, pride, and identity.

3 Smile: Camera-Based System

Smile facilitates the capturing and sharing of photos of people living with dementia in a long-term care environment. It consists of two components: (1) a camera in the communal space at the care facility and (2) a set of digital photo frames in the private rooms of the residents. Smile enables the person living with dementia to take a photo with visitors from outside the care environment or caregivers within and capture meaningful moments in an everyday care context.

The Smile camera is set on a tripod with a panel interface in front of the camera to control the system (see Fig. 1). Each button on the panel shows a photo, and the name of a participating resident in the care home (see Fig. 2). The user activates the camera by pressing the button with the resident's photo and name. The camera has a speaker for auditory feedback to guide users in taking a photo. It plays a voice message, asking the user to press the button a second time to confirm their intention and avoid accidental photos being taken. A voice message then explains that the photo will be taken in five seconds and counts down from five to zero before playing a shutter sound when taking a photo. During the design process of Smile, the main researcher compared responses to various types of cameras. In this study, people living with dementia were most familiar with the

Fig. 2 The panel with buttons. each button contains the photo and name of a participating resident to whom the photo could be sent

aesthetics of cameras prevalent in their formative years. Therefore, the Smile camera has a viewfinder, dials, flashlight, buttons, and lens to resemble a twentieth century camera style to make it recognizable for the residents. It was created using a laser cut box with a mini camera inside, controlled via Raspberry Pi.

Finally, the Smile system sends the photo to a digital photo frame in the private room of the residents with dementia whose photo was selected and provides voice feedback to indicate the completed process. The digital photo frames were placed in participants' private rooms as part of their everyday environment (Fig. 3). The digital frames displayed the captured photos of recent visitors along with the text: 'We visited you'. The photos serve as 'digital mementos' of a shared experience [21], family visits, or social activities in the care facility, that can be shared and talked about with family members and caregivers. For this study, we used off-the-shelf 10-inch digital photo frames that automatically displayed the photos taken at the Smile camera. The frames were set in 'slideshow' mode and also displayed the current date and time. In a previous iteration before this field study, the frames displayed a timestamp below each photo. However, this was confusing for the residents when shown alongside the clock showing the current time. As each new photo was captured at

Fig. 3 Digital photo frames located in the participating residents' private rooms display the photos taken with the smile system

the Smile camera, the digital frame issues a notification and displays the photo automatically. The user can then scroll through the photos using a remote control. The photo frame would go into sleep mode if no movement were detected in the room.

4 Method

We conducted a field study where Smile was deployed for four weeks in a care home for people with dementia located in the Netherlands to explore the role of Smile in supporting the relationships between the residents with dementia, caregivers, and relatives in a real-life care environment.

4.1 Setting and Participants

The study took place in a care unit in a care home where long-term care is provided for people in the advanced stages of dementia. The care unit offers residence to seven older adults living with severe dementia. Each resident has a private bedroom, bathroom facilities, and a communal living area and kitchen with around-the-clock care professionals available. Four residents with dementia were recruited to participate in this study, of whom three were female, and they were invited to use Smile freely during the study. All participants were in advanced stages of dementia, as confirmed by the care home. However, details of types of dementia, age, or symptoms were not requested. The main researcher (second author) was regularly present in the care facility to conduct observations and hold interviews with professional caregivers or relatives visiting the care facility.

4.2 Ethics and Privacy

The study was ethically approved by the care home's Client Representation Board. Family members of participants gave informed proxy consent to participate in the study and for their photo to be displayed on the Smile control panel at the camera in the living room. The participants with dementia were also asked to give consent, which involved the family member engaging with the resident to communicate information about the study and ensure they wanted to participate. The participants were also reminded that they could withdraw from the study at any time or request the deletion of captured photos. An informative poster was placed next to the camera system to provide information in-context on the study and the Smile system, in addition to the information letters. Furthermore, the professional caregivers were instructed to provide support and explanation in-person for visitors as requested.

Placing a camera in a complex and shared space such as a communal living area has ethical implications for privacy and the display of personal photos. The camera voice messages informed users that a photo was to be taken and actively asked participants to confirm they wanted to take a photo and display the photo in the private room of the selected receiver. No captured photos have been displayed in public communal spaces, only in the residents' private rooms.

Participants were free to use the Smile camera system how and when they wanted. While photos could be supportive and result in a positive experience, there was also the possibility that photos could be confronting or cause negative emotions in residents. Participants and their families were informed of this risk and advised that they could request the removal of a specific photo by contacting the main researcher. However, this was not necessary as no confronting experiences were reported during the study.

4.3 Data Collection and Analysis

The main researcher conducted observations on the use of the Smile system, including detailed descriptions of the context, interactions, and conversations [22]. After the field study, semi-structured interviews of approximately 10 min were conducted with two residents with dementia, three caregivers, and six visitors. The field notes and interviews have been analyzed using thematic analysis [23]. The qualitative data was analyzed in the native language by the main researcher through an inductive approach by identifying themes based on the quantity of occurrence and by clustering quotes relevant to the research question. The quotes were later translated into English for presentation in this paper.

5 Results

In total, 52 photos were taken by participants, together with five caregivers and 38 visitors. Six visitors used the photo camera during multiple visits, and approximately two-thirds of the photos were taken together with participants, as some visitors also took a picture without the participant before going to the room of their family member. The analysis of the field notes and interviews revealed three main themes and several suggestions for design improvements. Quotes are referenced as follows: resident with dementia (R), visitor (V), professional caregiver (C), and field note researcher (Obs#).

5.1 A Snapshot of the Everyday Life of the Resident

The photos offered relatives visiting the care facility insights into residents' social activities and everyday life in the care facility: "*It provides an impression of my mum's life here. Who visits [her], and how she presents herself*" (V1). Visiting family members felt curious who else visited and were positively surprised to see that other relatives had visited their loved one: "*Personally, I also like to see who visited. Some time ago, I heard from my brother that an aunt visited her. That is special—the number of people that are faithful and visit my mother. [...] But if my brother did not see her coincidentally, then we would not have known that she visited.*" (V6).

Visitors felt happy to see the resident's photos with professional caregivers, as it suggested they had a good relationship with the care staff and that they engaged in social activities: '*A visitor of a resident looks at the digital photo frame and sees a photo of a caregiver. The visitor asks the resident: "who is that caregiver? Never seen her...". The resident is silent and then mumbles. "She is very friendly"*' (Obs11).

The photos also informed professional caregivers about previous events and activities in the care home that the resident had participated in. By reflecting on those activities, the care staff could reconnect with residents during day-to-day care provision. A caregiver explained, "*if the person talks about a topic, in particular, I can maybe use the photos to trace back [to] what [activity or event] made that impact*" (C1).

It was unclear whether the photos made the residents remember or recognize visits or whether seeing the photo made residents understand that a visit had taken place. However, residents seemed to recognize familiar people in the photos and identify themselves: "*If I know the people on the photos, then I can tell you who they are. That is 1–0 for me*" (R1). Furthermore, the photos offered opportunities for reflection upon the residents' everyday life: '*The resident points to the photo of herself. She states that her photo is different from the other photos*' (Obs14). These examples illustrate how photos can offer insights into the resident's everyday life in a care facility and open up new perspectives for reflection and social interactions.

5.2 Enhancing Quality in Social Encounters

The photos enhanced quality in conversations by providing rich personal visual stimuli and triggered the residents to tell personal stories, as a caregiver explains: "*This morning I was in the room of Miss [surname] and had a glimpse at the photo frame. She noticed it and immediately shared that her old neighbor had visited her that day [...] And that is good as it helps [the caregiver] to get to know more about the person*" (C1). Another caregiver described how a displayed photo presented an opportunity to sit down and connect with the resident: "*A woman asked about who that girl was with brown hair and who has a funny way of saying 'hello'. So together with her, I looked at the photos on the screen to see whether it contained a photo of that girl*" (C3).

The photos created mutual understandings about the topic of conversations by providing visual context for interpretation: "*The visibility of who visited is good. You can grab the frame to talk about it. Without the photos, I would not manage to have her recall the person I am talking about*" (C1). A caregiver stated that the ease of access to the information is essential to be able to offer a personal approach during their routine care tasks, as the photos are part of the care environment: "*A daughter told me to take a look at it [the book with written records about activities and visits updated by the family]. However, that is not something I would do that easily because it is so personal. [...] A photo is something I can see in one glimpse. So I can ask something about it, like who the person is on the photo*" (C2).

Visitors and caregivers used the photos as visual proof of recent events and activities to reduce possible uncertainty and confusion experienced by the person living with dementia. Visitors used the photo as a motive for sharing information about the photographed person. One of the visitors explained: "*For the photo of [name grandchild] I, for example, tell how she is doing. Just the everyday information that I know. For example, that she is doing her final exams*" (V6). Caregivers used the photos during conversations with the resident to reduce stress or tension by showing visual proof, as the following example shows: "*One of the residents always asks when her husband will come to visit her, even if he has already come by. So at that moment, I took the photo frame, but unfortunately, they did not take a photo together. Otherwise, I could have easily shown her that he already visited her that day as she is wearing the same clothes as in the photo. If I just tell her, she would not believe me*" (C3). Another caregiver proposed that taking a photo could function as the last activity before saying goodbye for the visitor and resident: "*We just discussed the case that a husband of a resident living here does not want to leave, so he lingers at the door. In that case, we can say: 'let's take a photo to say goodbye'*" (C2).

5.3 Offering Dignity and Normalcy

Taking a photo was an exciting and social experience that provided a sense of normalcy to the lives of the residents with dementia. Visitors used the Smile system as a meaningful shared activity that offered opportunities to shift attention, give compliments, have physical touch, and include humor during their visit: '*A resident is wandering. A staff member is sweeping the floor and notices the resident. She stops her activity and states: "We have not yet taken a photo today together, come," and she waves her hand to indicate that the resident should approach her. Together they take a photo and have a small chat about it.*' (Obs27).

Being photographed also stimulated a sense of dignity and increased self-value for the residents. Taking a photo together allowed visitors to give personal care to the residents by, for example, doing their hair or dressing them up nicely. These actions resulted in small but private and intimate moments where the visitor could physically and socially connect with the resident: '*The visitor positions the resident in front of the camera and tells her to wait. The visitor walks fast-paced to the room of the resident and returns with a hairbrush. She brushes the hair of the resident and pulls her shirt down. As a response, the resident asks when the hairdresser will come*' (Obs6).

Professional caregivers also used Smile to provide a meaningful moment during everyday care for the resident to receive personal attention, compliments, and appreciation: '*Before taking a photo, the caregiver dresses up a resident and states that it looks perfect now. The resident seems happy and says thank you to the caregiver. The resident also fixes her clothing and moves hair out of her face. The caregiver asks: Are you ready?*' (Obs35).

Capturing and reviewing the photos was used by visitors and caregivers to give compliments to help maintain the resident's self-esteem and provide a feeling of normalcy and inclusiveness. Residents flourished from receiving comments about their appearance: "*You see that Miss [surname] feels honored when you ask her to take a photo, and you say that she looks good*" (C3). Furthermore, residents felt at ease and experienced a pleasant activity as they also made jokes about it: '*After taking a photo, a resident living with dementia says: 'Oh that is nice, so now I am a photo model?!' making all involved people laugh'* (Obs3).

Humor was commonly used by visitors and care professionals when using the camera, for example, taking photos while making funny faces. The caregivers also experienced exciting and joyful moments when seeing photos of colleagues joking and pulling silly faces. The caregiver mentioned: "*It is relaxing, a moment [that is] different than the daily care activities. It caused me to smile*" (C3).

5.4 Improvements for Design

The interviews and observations resulted in several suggestions for improving the design of Smile. For example, the person with dementia may lack understanding of the photo-taking activity: '*After the photo is taken, the resident asks why a photo has been taken and what she has to do now.*' (Obs15). Similarly, taking the photo can lead to confusion: '*The resident does not understand. "Who will take it [the photo]? Someone needs to look through the lens, right?"*' (Obs8). This suggests the residents with dementia need assistance from a caregiver or visitor to be involved in the activity. Smile was designed to be understandable and accessible, but the residents' abilities were very different. A visitor expressed the desire to have a step-by-step shared activity in which the pace could be adjusted to the situation: "*When I am explaining that we are taking a photo, the device is counting down, so halfway [through] my explanation I have to smile to the camera. My mother has no sense of what is happening at that moment*" (V6).

Furthermore, several residents looked at the control panel instead of into the camera lens, and many of the captured photos displayed residents looking down instead of straight into the camera. The audio message from the camera did not always get the user's attention, as there was no response to the audio instruction to look straight into the camera.

Some visitors were confused by the low quality of the photo, the subject in the photo, or the presentation. Some photos were too dark to see the content. Further confusion arose when several images displayed visitors of another resident as they were sent to the wrong person due to selecting the wrong photo on the control panel. Similarly, the automatic changing of the photos on the digital frames created some confusion: "*It is confusing that the photos are moving on [changing]. The image and story need to fit together*" (V6). Also, the order that the photos were displayed could be confusing: "*It confused me [a visitor] that the photos were displayed in a random order, or at least after the last photo, the first photo appears again. As a result, I have no clue about when somebody has visited. Maybe it was already weeks ago*" (V1).

A caregiver suggested a photo screen in the shared living room could allow photos to be on open display. But a family member noted that photos from other residents might not be familiar to other residents and possibly lead to confusion. Another option was for the personal photo frame to be expanded with more information regarding time, description of the people, or the occasion of the visit.

6 Discussion

Based on our results, we outline opportunities for camera-based systems such as Smile in long-term dementia care to support social relations and create moments for recognition of identity and dignity.

6.1 Strengthening Social Relations in Long-Term Care

Our results indicated that viewing photos of social events in the care facility informs visitors about the social environment and wellbeing of the resident with dementia. It can relieve feelings of guilt or lack of knowledge of day-to-day issues experienced by visiting relatives [2, 24]. Camera-based systems can support this by making activities of the resident's everyday life with dementia visible to relatives, who cannot be present at all times. For instance, relatives could gain insights into whom the person with dementia had socially engaged with throughout the day or the week.

Additionally, the straightforward and familiar act of taking a photo together can enhance connections between visitors or care staff and people living with dementia in formal care environments. Initiating and maintaining these relationships in close surroundings are essential to support a sense of 'self' or identity in people with dementia [25]. This sense of identity is expanded further in sharing photos with family members and care staff who comment and reflect on the person's photo. The photos facilitate new conversations and lead to storytelling and associations with personal aspects of everyday life by stimulating people to share memories. For instance, when people in the photo are recognized by the person with dementia, or when associations arise based on the photos presented in the frame. Therefore, capturing and sharing personal photos can positively influence self-esteem and support personhood [26], and enable person-centered approaches in dementia care [6].

6.2 Moments of Dignity and Recognition of Identity

Photos are often associated with special occasions, such as family gatherings or photo booths at marriages or graduations. After moving to a care facility, people with dementia often experience a lack of these special occasions. During the field study, it became apparent how preparing for a photo and taking the photo was a small but meaningful event in itself that provided a feeling of occasion within the everyday life of the resident. These events triggered family members and care staff to engage in a series of interpersonal actions that created and strengthened connections with the resident, such as preening, grooming, stroking, and touching, as shown in the comments presented.

Caregivers and relatives paid particular attention to the residents' appearance by dressing up, applying makeup, or brushing the resident's hair. Often, little attention is paid to the appearance of people with dementia in care homes, with the aging body being prone to stigmatization and marginalization [27]. Yet appearance is an essential aspect of how we self-identify and identify others as it becomes a signature of a person's sense of self and an expression of identity for people with dementia [28]. Therefore, there is an ongoing need for rituals and routines in the care environment to take care of and focus on appearance as part of everyday life [29]. We argue that capturing and sharing personal photos in dementia care environments can reinforce the residents' dignity and facilitate moments for self-identification and expression through choices related to appearance.

6.3 Limitations and Future Work

During the interviews, the participating people with dementia struggled to articulate their recollections about the act of taking a photo and the captured photos. Therefore, the in-context observations were valuable to ensure the first-person perspective of people with dementia was included in the evaluation. Further information is needed in this area, and further studies will carry out informal interviews during the use of the Smile system rather than reflecting on prior use. Future work will look at the engagement that occurs during taking a photo and the relationships that develop between residents and visitors. It will explore the ongoing design and use of tools to capture a photo with these visitors through refining and simplifying use and encouraging and enabling families to use their own devices to take photos and send them to the digital photo frame. The use of available off-the-shelf digital photo frames allows for accessibility and presents new opportunities for further inquiry and design exploration for integrating personal photos in long-term dementia care on a larger scale.

7 Conclusion

This paper presented and explored Smile: a camera-based system to increase visitor involvement and person-centered care engagement by professional caregivers with people living with dementia in long-term care. Our results reveal how camera-based technology can support social relations between professional staff, family members, or other visitors, and people living with dementia. The system contributed to the daily care practice by providing a shared activity in the care environment, contributing to personal contact and conversations, and providing a sense of normalcy and dignity for the residents. Visualizing the experiences and wellbeing of the residents living with dementia enabled visitors to overcomes the feeling of unfamiliarity and lack of insight into the day-to-day care environment. Also, Smile offered visitors the opportunity to shift attention away from a focus on dementia and engage with the person in a shared experience that acknowledged identity as revealed through a focus on appearance which resulted in attention, compliments, playfulness, and reinforced self-identity in the resident with dementia.

References

1. Kellett, U.M.: Transition in care: family carers' experience of nursing home placement. J. Adv. Nurs. **29**, 1474–1481 (1999)
2. Kenning, G., Treadaway, C.: Designing for dementia: iterative grief and transitional objects. Des. Issues **34**, 42–53 (2018)
3. Vines, J., Lindsay, S., Pritchard, G.W., Lie, M., Greathead, D., Olivier, P., Brittain, K.: Making family care work: dependence, privacy and remote home monitoring telecare systems. In: Proceedings of the 2013 ACM International Joint Conference on Pervasive and Ubiquitous Computing; Association for Computing Machinery, pp. 607–616. New York, NY, USA (2013)
4. Bramble, M., Moyle, W., McAllister, M.: Seeking connection: family care experiences following long-term dementia care placement. J. Clin. Nurs. 18, 3118–3125 (2009)
5. McNaney, R., Vines, J., Mercer, J., Mexter, L., Welsh, D., Young, T.: DemYouth: co-designing and enacting tools to support young people's engagement with people with dementia. In: Proceedings of the 2017 CHI Conference on Human Factors in Computing Systems—CHI '17, pp. 1313–1325. ACM Press, New York, NY, USA (2017)
6. Fazio, S., Pace, D., Flinner, J., Kallmyer, B.: The fundamentals of person-centered care for individuals with dementia. Gerontol. **58**, S10–S19 (2018)
7. Theurer, K., Mortenson, W.B., Stone, R., Suto, M., Timonen, V., Rozanova, J.: The need for a social revolution in residential care. J. Aging Stud. 35, 201–210 (2015)
8. Desai, A.K., Grossberg, G.T.: Recognition and management of behavioral disturbances in dementia. Prim. Care Companion J. Clin. Psychiatry **3**, 93–109 (2001)
9. Fredrickson, B.L.: Positive emotions broaden and build. In: Devine, P., Plant, A.B.T.-A. in E.S.P. (eds.) Advances in Experimental Social Psychology, pp. 1–53, vol. 47. Academic Press, ISBN 0065-2601 (2013)
10. Kim, C., Wu, B., Tanaka, E., Watanabe, T., Watanabe, K., Chen, W., Ito, S., Okumura, R., Arai, T., Anme, T.: Association between a change in social interaction and dementia among elderly people. Int. J. Gerontol. **10**, 76–80 (2016)
11. Houben, M., Brankaert, R., Bakker, S., Kenning, G., Bongers, I., Eggen, B.: The role of everyday sounds in advanced dementia care. In: Proceedings of the 2020 CHI Conference on Human Factors in Computing Systems, pp. 1–14. ACM Press, New York, NY, USA, Paper 450 (2020)
12. Frohlich, D.M., Corrigan-Kavanagh, E., Campbell, S., Chrysanthaki, T., Castro, P., Zaine, I., da Graça Campos Pimentel, M.: Assistive media for well-being. In: Brankaert, R., Kenning, G. (eds.) HCI and Design in the Context of Dementia, pp. 189–205. Springer International Publishing, Cham, ISBN 978-3-030-32835-1 (2020)
13. Lindley, S.E., Randall, D., Sharrock, W., Glancy, M., Smyth, N., Harper, R.: Narrative, memory and practice: tensions and choices in the use of a digital artefact. People Comput. XXIII Celebr. People Technol., 1–9 (2009)
14. Waycott, J., Davis, H., Vetere, F., Morgans, A., Gruner, A., Ozanne, E., Kulik, L.: Captioned photographs in psychosocial aged care: relationship building and boundary work. In: Proceedings of the 32nd Annual ACM Conference on Human Factors in Computing Systems—CHI '14, pp. 4167–4176. ACM Press, New York, NY, USA (2014)
15. Gaver, W., Boucher, A., Bowers, J., Blythe, M., Jarvis, N., Cameron, D., Kerridge, T., Wilkie, A., Phillips, R., Wright, P.: The photostroller: supporting diverse care home residents in engaging with the world. In: Proceedings of the SIGCHI Conference on

Human Factors in Computing Systems; Association for Computing Machinery, pp. 1757–1766. New York, NY, USA (2011)
16. Brewer, R.N., Gellner, M., Piper, A.M.: Portrait pigeon: an interactive photo messaging wall for seniors. In: Proceedings of the 2014 ACM International Joint Conference on Pervasive and Ubiquitous Computing: Adjunct Publication; Association for Computing Machinery, pp. 13–17. New York, NY, USA (2014)
17. Lazar, A., Thompson, H., Demiris, G.: A systematic review of the use of technology for reminiscence therapy. Health Educ. Behav. **41**, 51S–61S (2014)
18. Lazar, A., Edasis, C., Piper, A.M.: Supporting people with dementia in digital social sharing. In: Proceedings of the ACM Conference on Human Factors in Computing Systems—Proceedings, pp. 2149–2162. ACM Press, New York, NY, USA (2017)
19. Thoolen, M., Brankaert, R., Lu, Y. AmbientEcho: Exploring interactive media experiences in the context of residential dementia care. In: Proceedings of the 2020 ACM Designing Interactive Systems Conference; Association for Computing Machinery, pp. 1495–1508. New York, NY, USA (2020)
20. Piasek, P., Irving, K., Smeaton, A.F.: Case study in sensecam use as an intervention technology for early-stage dementia. Int. J. Comput. Heal. **1**, 304–319 (2012)
21. Bowen, S., Petrelli, D.: Remembering today tomorrow: exploring the human-centred design of digital mementos. Int. J. Hum. Comput. Stud. **69**, 324–337 (2011)
22. Emerson, R.M., Fretz, R.I., Shaw, L.L.: Participant observation and fieldnotes. Handb. Ethnogr, 352–368 (2001)
23. Braun, V., Clarke, V.: Using thematic analysis in psychology. Qual. Res. Psychol. **3**, 77–101 (2006)
24. Putnam, M., Pickard, J.G., Rodriguez, C., Shear, E.: Stakeholder perspectives on policies to support family caregivers of older adults with dementia. J. Fam. Soc. Work **13**, 173–190 (2010)
25. Wallace, J., Thieme, A., Wood, G., Schofield, G., Olivier, P.: Enabling self, intimacy and a sense of home in dementia: an enquiry into design in a hospital setting. In: Proceedings of the 2012 ACM Annual Conference on Human Factors in Computing Systems—CHI '12, p. 2629. ACM Press, New York, NY, USA (2012)
26. Astell, A.J.: Technology and personhood in dementia care. Qual. Ageing **7**, 15–25 (2006)
27. Ward, R., Holland, C.: 'If I look old, I will be treated old': hair and later-life image dilemmas. Ageing Soc. **31**, 288–307 (2011)
28. Campbell, S., Buse, C.E., Twigg, J., Keady, J., Ward, R.: Appearance matters: it's integral to our sense of self. J. Dement. Care **23**, 20–23 (2015)
29. Ward, R., Campbell, S.: Mixing methods to explore appearance in dementia care. Dementia **12**, 337–347 (2013)

An Immersive Art-Based Approach to Engaging with the Embodied, Sensory and Affective Experience of Living with Dementia

Gail Kenning, Jill Bennett, Volker Kuchelmeister, Natasha Ginnivan, and Melissa Neidorf

Abstract

This paper discusses the design and exhibition of *The Visit*, an immersive media artwork, featuring a digital character with dementia. Created from verbatim interviews with people living with dementia, the character 'Viv' shares experiences of audio and visual hallucinations, confusion and confabulation, as well as her insights and reflections. Viewers are thereby drawn into the perceptual world of dementia, but also challenged by the depth of the engagement with the character.

Keywords

Dementia · Immersive media · Empathy

1 Introduction

The Visit is an immersive experience which explores the relational dynamics between people living with dementia and others by staging an interaction between a realistic digital human character, 'Viv' and the viewer (visitor). The participatory artwork was produced as part of a research program investigating the lived experience of people living with dementia by an interdisciplinary team of artists, designers, technologists, psychologists and dementia researchers, who work closely with people living with dementia. The artwork focusses on the experiences of Viv, a woman with dementia who visitors interact with as she moves about her kitchen. The work draws the viewer into her emotional and perceptual world as she speaks both to the viewer and to a hallucinated child, as if both are seated at her table. The monologue is constructed near verbatim from interviews with, and reports of, four women living with (vascular) dementia who experienced auditory and visual hallucinations, confusion or confabulation, and shared details about their condition.

The artwork involved the creation of a realistic embodied character which presented a rich, contextual portrayal of dementia foregrounding the subjective lived experience rather than the pathology of the condition. Through Viv we gain insight into the agency and capacity of people with dementia as we see how she perceives,

G. Kenning (✉) · J. Bennett · V. Kuchelmeister · N. Ginnivan · M. Neidorf
fEEL (felt Experience and Empathy Lab), University of New South Wales, Sydney, NSW, Australia
e-mail: gail.kenning@unsw.edu.au

G. Kenning · J. Bennett
Ageing Futures Institute, University of New South Wales, Sydney, NSW, Australia

R. Brankaert et al. (eds.), *Dementia Lab 2021: Supporting Ability Through Design*, Design For Inclusion 2, https://doi.org/10.1007/978-3-030-70293-9_8

reflects upon and finds ways to manage the hallucinations that are part of her every day. This strengths-based rather than deficit focused characterisation [1] implicitly challenges preconceptions about dementia through a direct engagement in a compelling relational encounter. By focusing on the nature of interpersonal interactions and relational dynamics, *The Visit* creates conditions for empathy rather than merely sympathy [2].

This paper discusses the iterative process of development. The process began with semi and unstructured interviews of the participants' experiences. These were subject to close analysis by the research team to understand how they made meaning from them. Artists and designers then developed two immersive modes of artworks; a Virtual Reality (VR) headset and a life size screen as part of an installation which extended the onscreen space into the gallery environment through the use of props and an ambisonic sound track. The artwork was exhibited as part of an arts health festival. The two modes of viewing allowed for greater accessibility, the VR headset being more portable. In addition, it allowed for a focussed individual viewing.

2 Background

Understanding dementia from the perspective of those living with the condition is increasingly the focus of research and design [3]. However, all too often those living with the condition feel they are not understood, and their voices are not heard [4]. As one of the participants in this research program suggested "I remember when I was diagnosed [with dementia] it sticks firmly in my head how I was shocked that the so-called experts didn't know the reality. You know, the consultants couldn't tell me what to expect apart from, you know, the decline and the basics, but they didn't understand the—you know—how our senses are so affected and that was just totally alien to them and that shocked me" [PID 8]. This lack of knowledge and understanding of what it feels like to live with dementia can enhance perceptions of deficit and impede any recognition of intelligence, agency and other indicators of personhood [5]. Such perceptions and judgements that focus on deficits, can in turn impede any experience of empathy [6]. To overcome this, it is important to slow down the perception process and the rush to judgement of what it means to live with dementia.

Those with limited experience of interacting or being with people living with dementia are often confronted by not knowing what to do. There is a tendency to 'take-over' which can work against listening, being understanding, or just being *with* [7]. By cultivating empathy and greater understanding of inter-subjective (social and shared) dynamics, we can not only breakdown stigma, but facilitate inter-group understanding and motivate pro-social helping behaviors [8–10]. Empathy is a complex and multidimensional construct that Davis [11] suggests, "connects the responses of one individual to the experience of another". Empathy studies have explored this through the use of perspective taking or affective role playing [12]. It has also been explored using immersive technologies to simulate and promote imaginative perspective shifting.

Art-based approaches allow for embodied, sensory and affective modes of communication that can engender empathy and promote empathic interactions [13, 14]. Immersive technologies, in particular, have been shown to increase empathy and pro-social behaviours [15, 16]. Such immersive experiences focus on simulating physical context and promoting, 'being in the shoes of'. This has led to immersive media and VR in particular being posited as the "ultimate empathy machine" [17]. However, while placing people into a different space may prompt imaginative transposition to a new condition, it does not necessarily register the subjective presence or *lived experience* of the 'other'. In other words, temporarily occupying—or appropriating—the position of another may be illuminating but it is not the same as engaging with that other. The ability to imagine another's experience is one part of empathy—but interactive art may also develop the capacity for *being with* rather than simply simulating *being*. *The Visit* moves beyond the imaginative transposition, and the problematic politics of colonising the position of the other without any real person to person engagement. It facilitates the experience of "being with" another and the associated capacities of attunement, and deep listening that exemplify empathy in the classic person-centred psychotherapy advanced by Carl Rogers [6, 18] by attending to the affordances of an experience, as suggested by Gibson [19].

The Visit draws on the longer tradition of the 'talking head', screen-based monologue that breaks the fourth wall, addressing the user/viewer directly—an approach skilfully developed in the portraits created by Alan Bennett in his *Talking Heads* Television Series [20]. *The Visit* aims to cultivate, in the viewer, sensitivity towards personhood and agency by making them perceive themselves as part of a dynamic relationship in which the character of Viv, in this case, potentially knows more than the viewer.

The authenticity of the character of Viv, and her demonstrated agency and personhood, comes from the detailed testimonies and reports of four older female adults living with vascular dementia who were able to understand their conditions, and had strategies for managing their hallucinations, confusion and confabulations. The authenticity is further developed in the characterization in the use of professional actors to voice and develop the movements of the character portrayed.

2.1 The Immersive Experience

The Visit offers a content rich, immersive engagement running on two distinct platforms; a VR Oculus Quest headset and an installation consisting of a large screen on which the character is perceived as life-size, a chair—which matches the chairs seen onscreen to enhance the sense of presence,—and an ambisonic sound track accessed through personal headphones (see Fig. 1). *The Visit* is also being developed as an interactive real-time film on the apple and android App Store. In the installation version in the gallery, a viewer takes a seat, seemingly at her table, triggering the interactive sensors and causing Viv to become 'aware' of their presence. She begins offering them tea and food. A third place is set at the table which, it becomes apparent, is occupied by a hallucinated child who is included by Viv in the conversation. The encounter with Viv intentionally promotes ambiguity prompting the viewer to question if they too are part of Viv's hallucination. Through the unfolding conversation the confusion and clarity of Viv's world is revealed. She talks of her love of music, walks, and gardens, and her disagreements with various carers. She tells of her confusion at visual hallucinations, such as when her garden shed disappears, only to appear again later. The immersive spatial soundscape, heard through personalised headphones, underlines the emotional transitions and creates the impression for the viewer of being 'inside' Viv's hallucinated world. The viewer hears the sounds overhead of, Viv tells us, her father's typewriter clicking away "upstairs". Thus, the viewer experiences a little of Viv's perceptual world but then is reassured as Viv herself reveals that this is a hallucinated reality. The unfolding dialogue, revealing Viv as a complex character, aims to counter any preconceived judgement with regard to dementia. The immersive media creates a palpable and vivid sense of presence. However, the viewer is deliberately not presented with interactive choices in the manner of a video game, but the work is rather designed to operate on Viv's terms. The aesthetic elements combine to create conditions conducive to a relaxed and unhurried listening, supporting openness and attunement.

3 Methodology

3.1 The Participants

The interviews and reports of four women with vascular dementia were purposively selected based on the need for information rich cases (and no intention to aim for generalisability) [1]. Each of the women had audio and visual

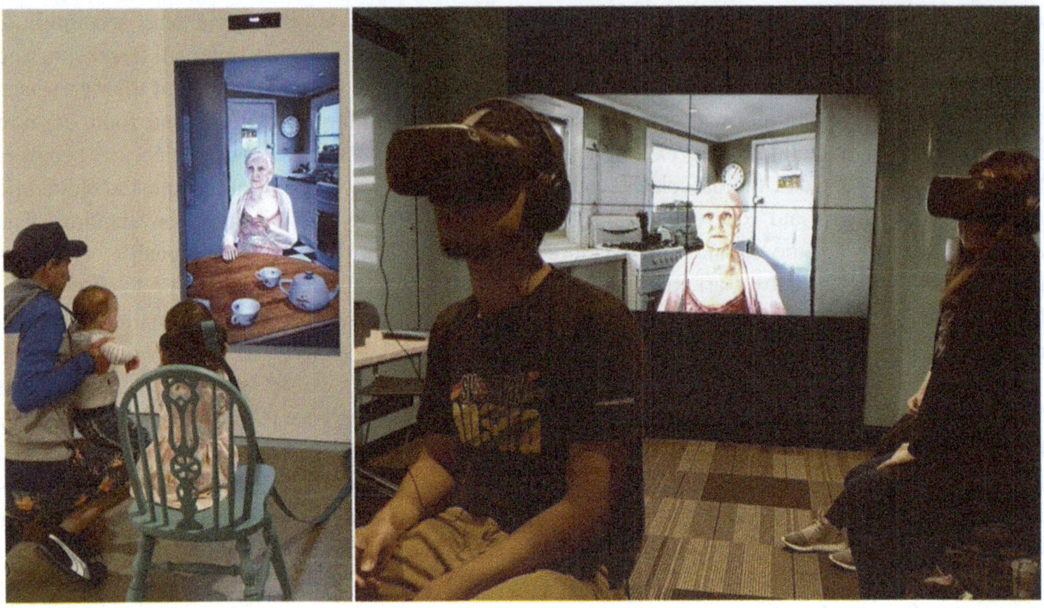

Fig. 1 *The Visit* left: the visit installation, [art festival name], 2019, [gallery name]. right: the visit VR [location]

hallucinations, confusion or confabulations and had an awareness and understanding of their conditions. Each were highly proficient in their chosen careers, with responsibilities, organisational, and virtuoso skills as General Practitioner (GP), health Administrator, musician, and writer and had raised families. We were able to gain insights into how their capabilities in their personal and professional lives had equipped them with coping mechanisms which were carried over into their 'after diagnosis' lives. They revealed the strategies they employed for adapting techniques from their parenting or professional lives and were able to reflect on both their auditory and visual hallucinations and on their experience of confabulation. They were able to provide information on how they had developed ways of 'testing' whether they were hallucinating or engaging with reality. They were also able to offer up ways of coping to their communities.

A series of interviews semi-structured and unstructured interviews were conducted over period of six months. Each participant was interviewed between three to eight times for between an hour and ninety minutes. The interviews were audio recorded, researchers made follow up notes, and the interviews were transcribed verbatim. The interviews began as semi-structured interviews focusing on their experience of ageing and changes that they had noticed to their cognitive and physical functioning. As the relationships and trust developed the interviews became less structured, allowing the interviewee greater control over the content discussed.

3.2 Experiential and Interpretative Analysis

The data was subject to analysis of phenomena relating to the experience of the participants and interpretative analysis of how they made meaning from these experiences [21]. Although, it did not employ Interpretative Phenomenological Analysis per se (IPA) [22]. The aim was to 'give voice' to the participants rather than representing them, and to move beyond 'first order' analysis which focuses on concerns and identifies themes [23]. Researchers were engaged in analysis of the data at an interpretive and conceptual level working from "highly intensive and detailed analysis of accounts produced by a comparatively small number of participants" [21]. The researchers' aims are to understand the participants' world, and 'what it is like' for them and to locate the initial 'description' in terms of wider social, cultural, and theoretical, contexts.

Researchers engaged with the interview material both individually and as a group. To 'hear' (and attune to) the voices of the participants, researchers listened to the audio recording of the interviews and in group analysis took turns reading the transcripts closely and out loud. This involved reading and re-reading and listening to the audio over again. The aim was to not only understand what was being said, but also how and why it was said. This means taking into account language used, expression and emotion and understanding how participants contextualised what they said in relation to their own background and experience of, for example being mother, GP, person with dementia, their coping strategy, and broader issues of for example, dementia as a social, cultural and medical phenomena in the twenty-first Century. The analysis revealed participants experiences of hallucinations, their life experiences which informed their coping strategies, their self-awareness of their condition, their ability to articulate their experiences, and how the re-telling of their experiences gave them agency and seemingly a sense of control over how they were understood and viewed.

3.3 Script Development

The phenomenological and interpretive analysis aimed at identifying the emotional valency of reported events and understanding the experiences of participants. In addition, a series of descriptive terms were applied to allow for an exploration of differences and similarities in storylines or experiences. They included, for example, awareness, cognitive, confabulation, coping, emotion, friends, hallucination, etc. The data was not thematized or categorized as such. A structuring element for the artwork was the inclusion of a piece of music discussed by one of the participants (and later voiced by Viv) by Australian composer Peter Sculthorpe. His string quartet no. 16 in five movements give rise to five 'chapters': I. Loneliness II. Anger III. Yearning IV. Trauma V. Freedom. A series of vignettes based on the interviews were created and allocated to chapters according to the five movements. The script was prepared for the immersive experience in a non-linear form, with storyline changing for each viewing of the *The Visit.* The script was developed with a core storyline and story options for each of the chapters (see Fig. 2). A team of experts in dementia, psychology, art, and design worked on the script in consultation with a scriptwriter. They selected topics for inclusion in the script that revealed both the agency of the participants, the impacts of dementia, and challenged the audience.

The creative process of developing script and dramaturgy from the verbatim and reported data meant that stories needed to be adapted to fit within a unified and a single character. This was done by selecting topics and stories that overlapped and allowed for continuity in the script. In addition, *The Visit* was staged in a domestic kitchen which enable Viv to be in her home, have agency over her surroundings and be in conversation with the viewer and the hallucinated child. To achieve this participants' data were combined, with additional scripting to link scenarios

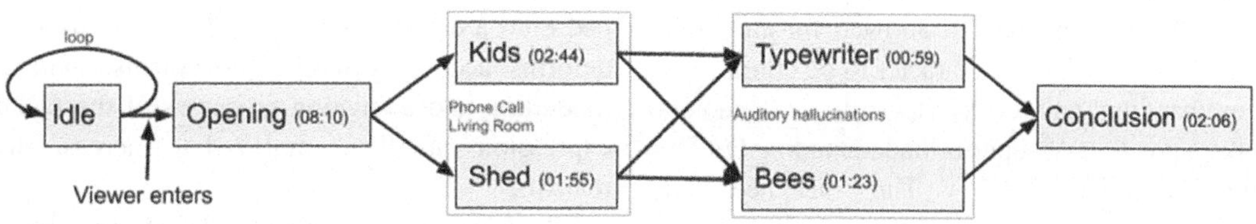

Fig. 2 The script developed into a non-linear narrative

and address the viewer. The script was developed in this way by the team in consultation with a professional playwright.

3.4 Character Construction

Viv is a composite portrait created from four women, whose accounts of living with mid-stage dementia (vascular dementia) were unified through the staged setting and the voice of an actor, who expressed the stories in the way they were related. The script was derived verbatim from semi-structured interviews, supplemented by additional reported activity. The aim was for all dialogue/events to have validity in terms of reflecting true experiences of vascular dementia. Minor modifications were made to the script detail to achieve consistency across the accounts, and names and identities of people mentioned were changed (the contributors themselves are credited with real names or pen names as per their individual preference). The script is thus intended to have scientific validity but is also driven by feeling and emotional valence, allowing for character development. The factual accuracy of Viv's monologue (her reality) is not always clear to the viewer, but she is compelling, smart, self-reliant and reflective. She experiences moments of confusion but is able to recover quickly. Viv has a strong, thoughtful, calm voice and a friendly but assertive manner. She sometimes gets 'on a roll' with her stories but in many places she pauses for reflection and reorientates herself.

A professional actor was employed to voice the character of Viv. The aim was to convey a range of controlled emotions, suggesting the inner emotional content of narrative that is sometimes broken and drifting. The voice which was developed for Viv had an authority and warmth to undercut the stereotypical notions of women with dementia. It allowed for ambiguity and depth and for the character to be alluring and commanding, and for the viewer to not be able to take a position of superior understanding but be a little intimidated by Viv. The voice needed to promote connection, encouraging visitors to sit with and listen to Viv without judgement, irritation, pity or despair, and to find a way into her world. While her monologue is coherent and more or less flowing it also has a slightly fragmented quality as one thought or memory leads to another and she seems to float off into her own reminiscing. She has a warmth and can be funny and wry, making knowing comments or little jokes. Viv makes eye contact and connects with the viewer, especially after 'drifting off topic'.

3.5 Virtual Production

While the data was analysed and the script was being developed, work began on the virtual production (VP) of *The Visit*. VP is a broad term used for computer-aided production and visualisation film-making methods. The process leverages real-time visualisation of characters and digital sets in combination with live-action capture of the cast. Here the physical elements of film-making are brought together with the digital of computer games technologies merging traditional practices with current and ongoing advances in real-time technology to enable filmmakers to make better creative choices earlier in the production process.

3.6 The Virtual Character

The virtual character of Viv was sourced as a 3D scanned digital human and adapted accordingly. Viv's movements were animated realistically by using Motion Capture (MoCap) to digitally record an experienced MoCap actor. MoCap is a process of digitally recording patterns of movement for the purpose of animating a digital 3D character in a film or video game. Unlike a video recording, a MoCap system records spatial data of the performers joints. This data is then mapped onto a digital human's skeleton and its skin deforms according to the movements, manual modelling and animation processes of the facial expressions are then employed to increase the realism.

The MoCap sessions were directed by one of the lead script writers [second author] and involved sessions in which the professional voice actor performed the character in situ. The MoCap actor then worked further with the director [second author] and immersive media designer [third author] to model the movement. Her voice was developed by well-known Australian TV and film actor Heather Mills who is experienced in film and television characterization. The voice was recorded separately with LipSync technology used to map the character's voice and lips. The character was depicted in a digital set which was a 3D scan of the 'real-life' functioning kitchen of a participant on the project (See Figs. 3 and 4 for of the virtual production processes and Fig. 5 of screenshots of Viv in her kitchen).

The artwork is designed to be viewed in VR on an Oculus Quest headset, which offers six degrees of untethered spatial tracking (not requiring a computer or cables). The viewer is only required to wear the headset and headphones and is not required to operate controllers. In addition, The Visit was exhibited onscreen in a gallery as an installation which was part of [exhibition name, Festival name and location]. Viv was perceived as life-sized in the interactive real-time video and was sensitive to the motions of the viewer, responding to the presence of the viewer sitting in front of the screen on the chair provided. The immersive experience is further supported by the ambisonic sounds of Viv's audio hallucinations and the kitchen environment, which is her reality are spatially perceived.

The viewer, seemingly seated on chair at Viv's kitchen table is required to adhere to the social conventions of visiting someone's home and is not free to roam around the virtual space. The call to movement and action of the viewer (the usual format of immersive experiences) is deliberately curtailed as the agency of the encounter remains with the onscreen character who 'recognizes' the presence of the viewer and moves and responds accordingly. Immersive experiences are often used to allow the viewer to 'step into' a character, in this artwork the awareness of the viewer becomes focused on their role *alongside* another. They are not positioned in this immersive experience, as 'seeing through the eyes' of Viv [15], which is often expected in a VR and immersive experience, rather they are encouraged to be *with* Viv.

4 Conclusion

The visit extends the field of experience simulation and relational VR because of its use of deep context rich data, and in positioning the viewer *beside* another, rather than offering a transpositional experience of becoming another.

Fig. 3 Visual production processes using motion capture

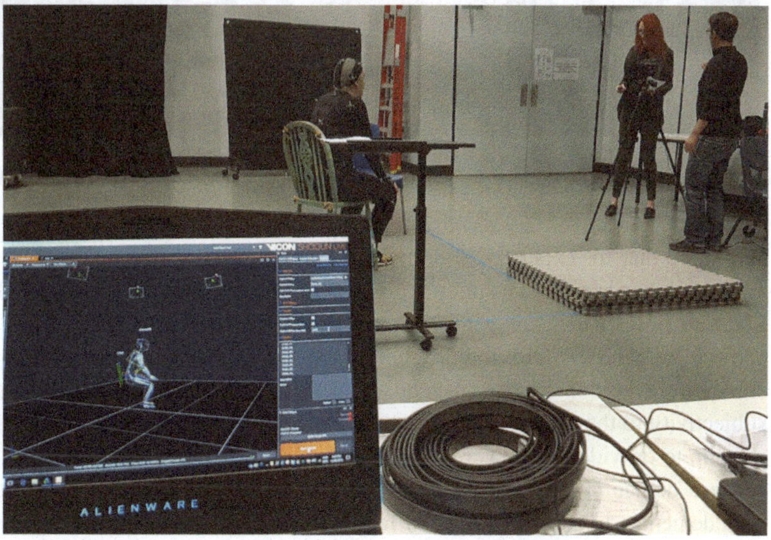

Fig. 4 Visual production processes using an actress to voice the character

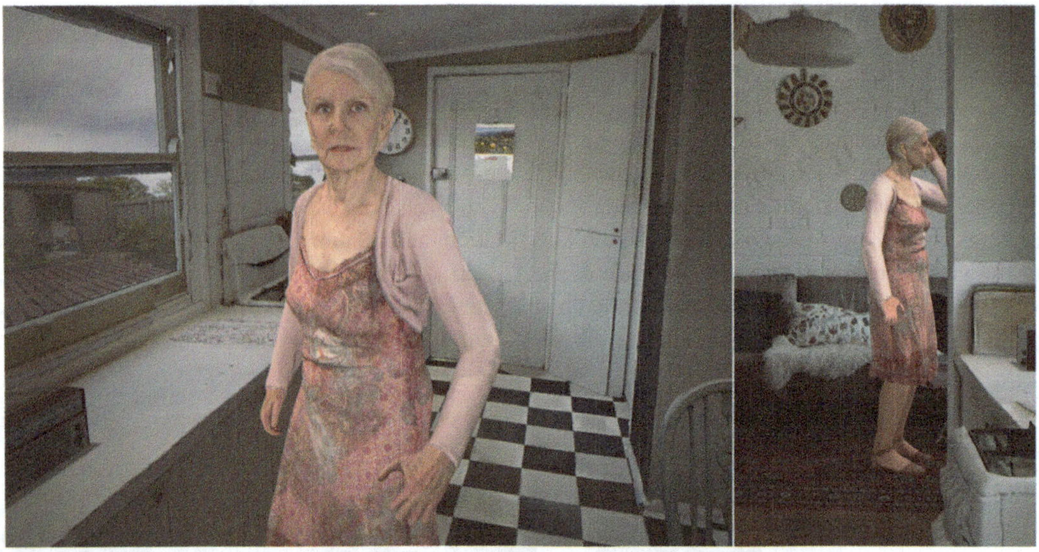

Fig. 5 Screenshots of the visit which show the digital character Viv in a 3D-scanned kitchen

We argue that to 'experiences homelessness' by virtue of being placed in a street at night as in the Stanford project or 'seeing through the eyes of another' or perceiving from a different embodied situation does not engage psychosocial inter-subjectivity to the same extent. The artwork does not aim to provide the viewer with what it is like to *have* dementia, but to be alongside a person with dementia, to cultivate empathy for what *they* are experiencing from being *alongside*. The script and character development employed the services of people experienced in dramaturgy and in how creative engagements can present nuanced and ambiguous experiences that avoid the didactic informational approaches of 'knowledge transfer'. In addition, the use of imagination and memory by the viewer ensures it remains a part of their knowledge base as an experience which promotes schema building and awareness of the issue. Viv, embodying the accounts of lived experience of people with dementia, both manifests the internal perceptual world of dementia *and* the capacity for insight and self-knowledge that confounds the judgement and preconceptions of dementia. A formal evaluation of the cultivation of empathy as a result of *The Visit* has also been published [in press].

It creates the potential for an experience by attending to affordances, rather than directing the experience per se [19]. It presents opportunities for action "that neither demand nor allow a participant to easily co-opt, normalise or resolve the experience of another" [14]. The non-linear narrative produced from assembled transcripts for use both on screen and VR give the viewer insights into the contested realities experienced by the character Viv and brought a greater awareness of what it is like to be *with* and *attuned to* a person facing these challenges.

The resulting psychosocial understandings are layered and complex as the psychological and social context of the participants are overlaid with the experience of the viewer and their engagement with an immersive media artwork exhibited in a gallery context, an arts festival, and in public and community space. As part of an international mental health arts festival, with film screenings in the US, and shown as part of health and wellbeing conferences the audience for *The Visit*, and psychosocial 'knowledge translation' of the lived experience of dementia is socially, demographically and geographically diverse.

References

1. Peacock, S., et al.: The positive aspects of the caregiving journey with dementia: using a strengths-based perspective to reveal opportunities. J. Appl. Gerontol. **29**(5), 640–659 (2010)
2. Sinclair, S. et al.: Sympathy, empathy, and compassion: a grounded theory study of palliative care patients' understandings, experiences, and preferences. Palliat. Med. **31**(5), 437–447 (2017)
3. Brankaert, R., Kenning, G.: HCI and design in the context of dementia. Cham, Springer International Publishing, Imprint, Springer (2020)
4. Villar, F., Serrat, R., Bravo-Segal, S.: Giving them a voice: challenges to narrative agency in people with dementia. Geriatr. (Switzerland) **4**(1) (2019)
5. Mitchell, G.J., Dupuis, S.L., Kontos, P.C.: Dementia discourse: from imposed suffering to knowing otherwise. J. Appl. Hemeneutics **12**, 1–19 (2013)
6. Rogers, C.R.: A theory of therapy, personality and interpersonal relationships as developed in the client-centered framework. In: Koch, S. (ed.) Psychology: A Study of a Science. Study 1, Volume 3: Formulations of the Person and the Social Context, pp. 184–256. McGraw-Hill, New York (1959)
7. Phillipson, L. et al.: Dementia attitudes and help-seeking intentions: an investigation of responses to two scenarios of an experience of the early signs of dementia. Aging & Ment. Health **19**(11), 968–977 (2015)
8. Batson, C.D. et al.: Empathy and attitudes: can feeling for a member of a stigmatized group improve feelings toward the group? J. Pers. Soc. Psychol. **72** (1), 105–118 (1997)
9. Batson, C.D. et al.: Five studies testing two new Egoistic alternatives to the empathy—altruism hypothesis. J. Pers. Soc. Psychol. **55**(1), 52–77 (1988)
10. Ahn, S.J., Le, A.M.T., Bailenson, J.: The effect of embodied experiences on self-other merging, attitude, and helping behavior. Media Psychol. **16**(1), 7–38 (2013)
11. H., D.M.: Empathy. In: Stets, J.E., Turner, J.H. (eds.) Handbook of the Sociology of Emotions. Springer, Boston (2006)
12. Galinsky, A.D., Moskowitz, G.B.: Perspective-taking: decreasing stereotype expression, stereotype accessibility, and in-group favoritism. J. Pers. Soc. Psychol. **78**(4), 708–724 (2000)
13. Bennett, J.: Empathic vision: affect, trauma, and contemporary art. Stanford, CA, Stanford University Press (2005)
14. Bennett, J., Froggett, L., Muller, L.: Psychosocial aesthetics and the art of lived experience. J. Psychosoc. Stud. **12**(1) (2019)
15. Herrera, F. et al.: Building long-term empathy: a large-scale comparison of traditional and virtual reality perspective-taking. PLoS ONE. **13**(10)
16. Oh, S. et al.: Virtually old: embodied perspective taking and the reduction of ageism under threat. Comput. Hum. Behav. **60**(C), 398–410 (2016)
17. Milk, C.: How virtual reality can create the ultimate empathy machine. TED, Ideas worth spreading (2015)
18. Rogers, C.R.: The necessary and sufficient conditions of therapeutic personality change. J. Consult. Clin. Psychol. **60**(6), 827–832 (1992)

19. Gibson, J.J.: The ecological approach to visual perception. New York, NY, Psychology Press (1986)
20. Bennetts, A.W.: Talking heads. In: S.D. Burge (ed.) UK (1988)
21. Larkin, M., Watts, S., Clifton, E.: Giving voice and making sense in interpretative phenomenological analysis. Qual. Res. Psychol. **3**(2), 102–120 (2006)
22. Smith, J.A., Flowers, P., Larkin, M.: Interpretative phenomenological analysis: theory method and research. London, Sage (2009)
23. Tuffour, I.: A critical overview of interpretative phenomenological analysis: a contemporary qualitative research approach. J. Healthc. Commun. **2**(52), 1–5 (2017)

Assembling a 'We': Tracing the History of My Doctoral Research to the People Who Helped Shape It

Christine Novy and Roanne Thomas

Abstract

In this paper I (first author) discuss my creative collaboration with two participants from the *Life Story Project*, a performative life story initiative which I facilitated at a day program for people with dementia in Ottawa. Drawing from journal and photographic records, I highlight what I learned from these two storytellers about communication. In particular, I reflect on the value of multi-modal communication methods and media, such as pictures, sounds and embodiment, as means to bypass word-finding difficulties, access embodied knowledge and negotiate meaning. I then discuss how these understandings helped shape the design of my current doctoral research, *Linking Lives Through Care: A narrative inquiry into relational care understandings*, which explores the potential for a more inclusive research environment in a long-term care setting. Tracing the history of my doctoral research to the people who helped shape it has been a way for me to keep their influence alive but, equally, to design a research protocol not only *for*, but, more importantly, *with* people with dementia.

Keywords

Dementia · Design · Inclusion · Research

1 Introduction

> Critical qualitative inquiry is, I would argue, invested in assembling a 'we.' While other research modes and practices are rightly and very capably interested in describing the world (a collective), our scholarship is—and should be—invested in gathering people together to create an us: one that brings about a plurality that is invested in one another. [1]

This paper is about inclusion, about creating a 'we', on a number of different levels. At its centre are two storytellers, Maria and Antonella (pseudonyms). Both are living with dementia and both were participants in the *Life Story Project*, a performative life story initiative facilitated by the

C. Novy (✉) · R. Thomas
University of Ottawa, Ottawa, Canada
e-mail: cnovy068@uottawa.ca

R. Thomas
e-mail: roanne.thomas@uottawa.ca

R. Brankaert et al. (eds.), *Dementia Lab 2021: Supporting Ability Through Design*, Design For Inclusion 2, https://doi.org/10.1007/978-3-030-70293-9_9

first author and hosted by a nonprofit day program in Ottawa. Throughout the *Life Story Project*, I (first author)[1] kept a journal in which I reflected on the methods used and their effectiveness. A photographer was also engaged to document the life story performances. Drawing from these records, I discuss how narrative and the dramatic arts were used to work around the communication difficulties experienced by Maria and Antonella to promote experiences of inclusion. The *Life Story Project* led to my current doctoral research, *Linking Lives Through Care: A narrative inquiry into relational care understandings*, which explores the potential for a more inclusive research environment in a long-term care setting. As a new researcher, I am aware that writing and language can work to include, but, equally, to exclude people with dementia. My background in narrative therapy and community work has taught me, however, that stories about who we are and who we have been are interwoven with our relationships [2]. My creative encounters with Maria and Antonella stand out as turning points towards greater understanding. As such, they contributed significantly to the development of my ideas. In this paper I bring together narrative elements that allow for a contextual interweaving of their contributions with reflective statements of my own grounded in academic literature. By tracing the history of my research design to the people that helped shape it, I hope to make their contributions more visible and keep their influence alive. I begin by positioning myself and my interest in arts-based communication. I then provide an overview of the *Life Story Project* history, objectives and process. Following which, I discuss my encounters with each of the two storytellers and the insights that arose from our work together. Lastly, I discuss how these insights were used to design a more inclusive communication environment in the *Linking Lives Through Care* study.

I (CN) am a drama therapist by profession and, formerly, I have worked with marginalized groups: children, youth and women who have experienced complex trauma. In dramatherapy, dramatic media such as movement and props are used to give external expression to inwardly held feelings and experiences [3]. In situations where an experience is simply too painful or complex to talk about these media offer an alternative language and a way 'to speak from inside' [4, 5]. When I began my work at a day program for people living with dementia, I was curious to explore whether these same dramatic languages might be useful in building bridges of communication between the program participants and members of their community. At that time, I began reading in the field of dementia studies and was encouraged to find researchers, who endorsed the use of non-verbal, embodied communication as a means for people living with dementia to initiate and sustain connection with others [6–10].

2 The *Life Story Project*

The life story project took place at a nonprofit adult day centre providing activity programmes for people living with dementia in the community. I began my relationship with the organization as a volunteer. This involved assisting with activities such as meet-and-greet, sing-alongs and individual accompaniment and provided me with an opportunity to learn about dementia, its effects and how best to support people living with dementia. In time the organization applied for funding to support my work and the *Life Story Project* was born. Referrals to the project were made by the program team leader and were based on her appraisal of whether a person might enjoy participating. Initially, the recruitment process was fairly informal. I would sit with the person in question, explain my interest in stories and ask if there was a particular story from their past that they would be happy to share with members of their community. Often, without further prompting, they would begin to tell me about a special person, place or time in their life. The

[1]Author[1] prepared the manuscript, which is largely based on her practice before taking up doctoral studies. Author[2] supervised Author[1]'s doctoral study *Linking Lives Through Care* and advised on manuscript drafts.

program team leader would then seek consent from their family member/Authorized Decision Maker. In time, and, as family members became more involved in the project, recruitment meetings would also include the prospective storyteller's family member/s and written informed consent would be sought at that time.

The *Life Story Project* was designed for people for whom communication had become increasingly difficult. We live in a verbal-centric society where language creates access to the social world [11]. While every effort was made to create an inclusive environment at the day program, an emphasis on verbal and written language during group activities left some participants on the margins, with fewer opportunities for self-expression or social interaction. My role at the day program was to work alongside these individuals and, with their consent, to record stories from their lives about events, people and places that were meaningful to them and that might be shared with others. As I did so, I looked for ways around their difficulties. My background in the dramatic arts had equipped me with a receptiveness to non-verbal communication and a set of skills that could be used to extend the expression of stories beyond words. My involvement in narrative community work had taught me the value of creating forums where these alternative stories might be witnessed and validated.

2.1 The Life Story Process

The life story process involved two distinct phases of work. During the first phase, I met with each storyteller individually over a number of sessions to record a story from their lives. These sessions took place in a quiet location aside from the main program activities. At the first session I would have on hand various storytelling resources inspired by our initial meeting. These might include a book of maps in which to locate their story, photographs loaned by their family, if these were available, as well as images downloaded from the internet to provide some cultural and historical context for their story. The second phase of work involved an interactive performance of their story in a group setting. Audiences typically included members of the storyteller's family, programme staff, and program participants, many of whom were themselves dealing with a range of cognitive and communication impairments.

To prepare for these gatherings, I would draft a life story performance script that combined the storyteller's own words with visual, auditory, and embodied storytelling. Non-verbal methods and media might include: (i) bold signs, large maps and projected images to indicate the story's location; (ii) objects featured in the story that could be passed around, handled or worn; (iii) short reenactments of key events in the story; and (iv) sound and music to evoke a particular era. In Antonella's Story, for example, we passed around produce—olives, potatoes and tomatoes—grown and harvested on the farm in Italy where she was raised; "women in the audience were invited to wear headscarves in the fashion of rural Italian matriarchs and, to punctuate our storytelling, Antonella's husband played some traditional Italian folk music on his accordion" (p. 33) [12]. Seating was arranged in a semicircle to ensure maximum visibility and staff and volunteers placed themselves beside participants needing extra support. This collaborative team approach, combined with a highly visual and interactive method of storytelling, made it possible for everyone present "to follow, engage with and themselves contribute to the story being told" (p. 99) [13].

2.2 Ethical Considerations

To accommodate memory loss, each storyteller's assent to participate was renewed at each new phase of the project. This process approach to consent [14] involved them in joint decisions as we went along. At our first meeting, for example, I would ask if they would like me to record their story so that it might, potentially, be shared with others. Once their story was complete and approved by the storyteller, I would again ask if they would like to share their story. This time I would explain the process of performing their story in front of an audience and ask whether they would like to invite members of their family.

3 Maria

Over its three-year run, 20 storytellers participated in the *Life Story Project*. Maria was one of the first participants and, at the time of our meeting, she was struggling with expressive aphasia, a condition that causes severe word-finding difficulty. Our collaboration began with a phase of individual work over several weeks. This was followed by an interactive performance of her story at the day centre. I have written about the performance of her story in detail elsewhere [13]. Herein, I reflect more closely on the collaborative ways in which her story was crafted and how our process of collaboration informed my work going forward.

3.1 Maria's Wartime Story

When I first met Maria and she began to speak about her past, I understood very little. She mentioned something about the war and about traveling by boat, but her words were jumbled and she frequently lapsed into her mother tongue of Dutch. At our second meeting, I brought with me a map of the Netherlands. This enabled Maria to locate her story and, using a mix of body language, Dutch and English, to trace the journey she had made as a child from Amsterdam to a province in the north of the Netherlands. This time, with Maria's permission, I wrote down her words and, although there were gaps in my understanding, I prepared a transcript of her story. As I did so, I tried, as far as possible, to reproduce her language and idioms. If a preposition or article was missing, I included what I thought made sense in brackets. If the content of her story was unclear or had been spoken in Dutch, I searched for images on the internet that might assist Maria to clarify her meaning and brought these, along with the transcript of her story, to our next meeting.

Listening to her story read aloud seemed to prompt Maria to supply lost words and, together, we began to fill in some of the gaps. She informed me, for example, that the year was 1944, and she remembered the urgency of her father's voice as he requested she be taken on board the boat. Thereon, our work together became more collaborative. Besides words, our negotiations involved pictures and sounds. At our third meeting, for example, I brought several pictures of boats downloaded from the internet and asked Maria whether any of these resembled the boat she had travelled in. While none was a perfect match, she used the pictures to differentiate the boat in her story. It was flat, carrying coal, covered and without insignia. On another occasion I brought auditory prompts and Maria used them in a similar way. When I played the sound of a diesel-engine canal barge, for example, she told me that the volume was too high, that the engine had barely been audible. Thus, little by little, in these collaborative and creative ways, we pieced together her story.

When a person is unable to communicate in a conventional manner, Bartlett and O'Connor [15] have written, it is easy to assume they have little to communicate. My biggest lessons as I worked alongside Maria were to stay open, to look for ways around her communication difficulties and expect to be surprised. Mid-way through our work together, for example, I learned that while Maria had difficulty articulating her thoughts, she was able to read the transcripts of her story with fluency. This made it possible to extend our collaboration into the performance of her story. My task was to bring Maria's story to life using a mix of visual, auditory and embodied storytelling; while Maria, at opportune moments, agreed to stand and read aloud an extract from her story. Thus, Maria was able to share her story with members of her community in a form that was recognizable and to assume what Baldwin [6] has called her 'narrative agency'.

Narrative Agency. As I reflect on what made this possible, I see how Maria's responses to the pictures, sounds and transcripts that I brought to our meetings alerted me to what was helpful and guided my subsequent input. In this way, facilitation—"enabling a person to do what otherwise he or she would not be able to do, by providing those parts of the action - and only those - that are missing"—merged into negotiation and collaboration (p. 91) [16]. In his description of positive person work, Kitwood [16] defines the process of interaction in three stages: (i) an individual makes an action; (ii) a second individual interprets the action and responds; and (iii) the first individual then interprets the response and reflects on whether the action was successful. Kitwood's definition allows for a mutuality of contribution in which the agency of both people is recognized. In the context of dementia care, care giving has more often been viewed as unidirectional and the agency of people with dementia, particularly those for whom verbal communication is difficult, has largely been discounted [15]. Yet, increasingly, researchers in this field recognize non-verbal communication—facial expression, expressive gestures, sound and tone of voice—as meaning-making resources, and embodiment as a source of knowledge [8–10]. Maria's contributions to my understanding were largely non-verbal. She showed me, for example, that pictures, sound and gesture helped to alleviate her word-finding difficulties. She taught me, through our negotiations, how important it was to represent her experience authentically and, when the time came to share her story, she stepped out from the shadow of dementia to be seen on her own terms. These understandings gleaned from Maria's responses and contributions to our work together informed my subsequent work with Antonella.

4 Antonella

4.1 Antonella's Family Story

Antonella's life story performance took place towards the end of the *Life Story Project* and by this time the lessons learned from Maria were fully integrated into the process. Antonella was born and raised in Italy and, at the time of her participation in the project, had lost most of her English language communication skills. Thus, we began our time together looking at pictures. Two had been supplied by her family: a passport photograph of Antonella at the time of her emigration to Canada; and a photograph of two small boys, her two sons, playing in front of their family home. The third picture, which I had downloaded from the internet, was of the small medieval town in Italy where she was born and raised. With these three pictures in front of us, Antonella told me, in just a few short sentences, about leaving Italy to begin a new life in Canada. There she met and married a compatriot and stepped into the role of mother to his newborn son. I have written about the performance of Antonella's story in detail elsewhere [12]. In this account, my objective is to highlight features of her life story performance, and Antonella's contribution in particular, that changed my thinking and inspired my current research.

A Visual Record. During the life story performances, a photographer was engaged to capture in-the-moment embodied responses and interactions. Among the photographs taken during Antonella's life story performance, one, in particular, stands out. It was taken towards the end of the performance, as we introduced Antonella's family. Just prior, I had read aloud the final words from Antonella's story: "When I took Marco he was a baby, I took care of the baby" and, as I did so, my colleague approached Antonella cradling a life-like baby doll wrapped in a soft blue blanket. The photograph was taken at the moment she received the doll into her arms. As she did so, Antonella transitioned from her audience role to become an actor in her story. Enrolled as her younger self, she was now able to show rather than to tell how she took care of the baby. A closer look at the photograph reveals Antonella's arms to be slightly blurred, which seems to suggest a rocking movement. Her gaze is downwards towards the face of the doll, and her mouth is slightly open, as though she might be speaking or singing to a baby.

Performance as a Way of Knowing. Life story performances were social occasions and designed to be interactive. Attendees at Antonella's performance included her husband and eldest son, staff and a number of Antonella's peers at the day program. As mentioned in the introduction, to ensure that everyone attending was able to follow and engage with her story, a variety of non-verbal storytelling methods and media were used to reinforce and frequently replace verbal narration. My collaboration with Maria revealed how these alternative communication methods can work to allay word finding difficulties and establish conversational common ground [17]. My account of Antonella's life story performance, above, provides further illustration of this. Schweitzer and Bruce [18] explain how familiar objects from the past are used in reminiscence work to trigger procedural memory: "as the hand feels the weight of an object, its texture, the eye sees its shape and the body replays associated movements" (p. 30). Thus, in the context of her life story performance, the baby doll prop enabled Antonella to bypass her verbal language difficulties and, in an embodied way, to add detail and significance to her story.

What stood out most about the photograph, however, was the quality of care that Antonella demonstrated: the attentiveness of her gaze and the competence of her holding. Kontos [10] was among the first to explore embodiment as a source of care knowledge in dementia research. Her ethnographic study in a long-term care setting reveals how gesture, facial expression and touch were used by the residents she observed to convey affection, offer comfort, and express support. In the photograph, as Antonella evoked the presence of a newborn baby, we see similar evidence of the care knowledge and care giving skills that she brought to her role as a young mother. Thus, much like Maria's, Antonella's contributions to my understanding were largely non-verbal. She showed me, for example, that the body is a site of knowledge [19] and that the dramatic arts could be used to make this knowledge visible.

Narrative Interdependency. In the same photograph, Antonella's eldest son, Marco, is seated beside her. While Antonella held the baby doll, Marco is shown responding to her story. He is clearly moved. The placing of these two care givers and their inter-connected stories side by side provides a striking example of what Baldwin has called 'narrative interdependency' [20]. In the same frame we see Antonella caring for Marco as a newborn baby, at the start of their relationship, and beside her we see Marco, now an adult with a family of his own, caring for his mother with dementia. By highlighting the 'mutuality of contribution' that exists between those who give and receive care, this photograph would seem to offer an alternative story about care; one that reflects the fluid and complex interdependencies that exist within families and across communities [21, 22]. As such, it inspired the title for my current research, *Linking Lives Through Care*, and became a source of inspiration for its design.

5 My Current Research: The *Linking Lives Through Care* Study

The *Linking Lives Through Care* study is a performance-based narrative inquiry that will take place in a long-term care setting and will bring together all three members of the care triad —residents who are living with dementia, family members and personal support workers—to explore relational care from different positions and perspectives.

5.1 Background to the Study

Studies that centre the experiences of those who receive care are few in comparison with studies representing professional or decision-making agendas. This is especially true for people who are cognitively frail and residing in long-term care settings, whose experiences of care have largely been marginalized [23, 24]. In these settings, care giving is viewed, above all, as a task-based service and residents are more often defined by their dependency on care [25, 26]. Yet, according to Tronto [27], care is a part of human life and, as such, is interwoven with "relationships of receiving and giving over a lifetime". This was particularly noticeable during Antonella's life story performance, as mother and son were united in their past and present caring relationship [12]. Tracing the contributions made by those cast in the receiving role to the lives of others can help shift the balance of power and, potentially, give rise to more equal and inclusive relationships between people with dementia and their partners in care [20]. Yet vulnerability does not preclude reciprocity [21]. As Kontos [10] has shown, people with dementia who reside in long-term care settings continue to care for one another through verbal and non-verbal exchanges. Brown Wilson and Davies [28] found, similarly, that residents, along with their families, do contribute to the process of care. However, unless these contributions are recognized, they can have little influence on the care provided.

The *Linking Lives Through Care* study is designed to re-centre the voices of people with dementia in research that concerns them and, in this way, give voice to and make visible their contributions to the process of care. A full account of the methodological considerations that will inform the study has been submitted for publication. In this section of our paper, I discuss how my encounters with Maria and Antonella have helped shape the study design. In particular, I focus on aspects of the methodology intended to create a more inclusive communication environment and, thereby, enable the participation of people with dementia who may find it difficult to express their views using just words.

5.2 A More Inclusive Communication Environment

In research, Hubbard, Downs and Tester [29] suggest, biographical stories offer some people with dementia an avenue through which "to give their experiences a voice" (p. 360). However, while some people with dementia are able to tell stories about their lives independently, others, like Maria and Antonella, can do so only with support [30]. During my work with these two storytellers they revealed to me the kind of support that is helpful. From Maria, for example, I learned the value of visual and audio props in helping to alleviate her word-finding difficulties and create conversational common ground between us. While these understandings mostly arose during the recording phase of our work together, Antonella's contributions to my understanding came to light during her life story performance itself. In this context, the invitation to show rather than to tell about her experience enabled Antonella to expand upon her story not through words but through actions. Thus I learned from both storytellers that a multi-modal approach to storytelling that combines verbal with non-verbal modes of communication, such as embodiment, visual and sound media, can give a voice to their experiences and open up new channels for dialogue.

Hydén [30] uses the term scaffolding to describe the advance planning that is necessary in order for those with linguistic and cognitive disabilities to access shared meaning-making resources. Beside Hydén, a number of narrative scholars in dementia studies underline the importance of accompaniment, collaboration and/or active facilitation to validate remaining communication abilities and support their continued use [6], [7], [31]. Maria and Antonella's indications about the kinds of support they required in order to participate on a more equal footing are echoed by practitioners and researchers alike. In a research context, for example, Hubbard, Downs and Tester [29] noticed how participants with memory impairments more easily engage with and reflect on 'here-and-now' experiences or actions that are directly observable. In a practice context, Schweitzer and Bruce [18] have argued, similarly, that recognition is easier than recall. These ideas have been incorporated into innovative visual and sensory approaches to research. Ward and Campbell's [32] 'Hair and Care' project, for example, involved in situ video-recorded conversations with people living with dementia during their visits to care-based hair salons. Buse and Twigg's [33] 'wardrobe interviews' involved interviewing people with dementia "alongside their wardrobes" (p. 1119); while Hung et al.

[34] used “go-along interviews” to ascertain the views of patients living with dementia about the hospital environment. In the context of conversational storytelling, Hydén [9] recommends, similarly, moving beyond the understanding of narrative as text ‘to the enactment and performance of the story’ (p. 365). In each of these examples, a multi-modal approach to communication, one that privileges sensory and embodied perception, is used to give a voice to the experiences of participants for whom verbal conversation has become increasingly difficult.

5.3 Different Ways of Knowing

If research is to be genuinely inclusive, Holmes, Murray, Perron and Rail [35] insist, then it must embrace alternative forms of knowledge and knowledge production. During my early collaboration with Maria, I learned to recognize her use of gesture, tone of voice and facial expression as meaning-making resources [30]. Later on I saw how these same meaning-making resources were used by Antonella to contribute in an embodied way to the retelling of her story. From both I learned that multi-modal storytelling methods and media helped validate their existing communication abilities, but, equally, they offered new routes to their interior worlds and alternative ways of knowing [36].

The *Linking Lives Through Care* methodology will involve a series of knowledge building sessions to allow for different retellings of the same stories and to accommodate different ways of knowing: verbal, embodied and interactive. The sequencing of these sessions will ensure that storytelling support is built into the process. To this end, I will be joined by a small team of artist collaborators who will assist in bringing each of the stories to life. During the interactive phase of knowledge-building, a videographer will also be present. This will ensure that both verbal and non-verbal contributions and responses are represented in the data. It will also provide information about the process of knowledge co-construction itself. For example: (i) how the stories about care giving and receiving are co-constructed through interaction and dialogue; (ii) the extent to which multi-modal storytelling methods and media can assist in accessing embodied knowledge and contribute to making it visible; and (iii) whether and how these same methods and media enable participants with dementia to give their experiences a voice.

In these different ways, the *Linking Lives Through Care* study is designed to be more encompassing of marginalized knowledge and forms of knowing. We are hopeful this will contribute to a more democratic and inclusive communication environment for the participants with dementia and thus a more participatory and collaborative knowledge-building process.

6 Conclusion

At the start of this paper, we included a quotation from Holman Jones [1] in which he likens qualitative inquiry to building community: “gathering people together to create an us”. In this paper, I (first author) have traced the contributions made by Maria and Antonella, both of whom are living with dementia, to my practice and research. Recognizing these contributions, making them more visible and acknowledging their influence has been a way for me to create an ‘us’ and, thereby, to design a research protocol not *for*, but *with* people with dementia.

References

1. Holman Jones, S.: Assembling a we in critical qualitative inquiry. In: Denzin, N.K., Giardina, M. D. (eds.) Qualitative Inquiry in Neoliberal Times, pp. 130–35. Routledge, New York (2017)
2. Combs, G., Freedman, J.: Narrative therapy’s relational understanding of identity. Fam. Process. **55**(2), 211–224 (2016). https://doi.org/10.1111/famp.12216
3. Jones, P.: Drama as therapy volume 1. theory, practice & research (2nd Ed.). Routledge, Howe (2007)
4. Anderson-Warren, M., Grainger, R.: Practical approaches to drama therapy: the shield of Perseus. Jessica Kingsley, London (2000)
5. Novy, C.: The narratives of change project: dramatherapy and women in conflict with the law. In: Jones, P. (ed.) Drama as therapy volume 2. clinical

work and research in practice, pp. 67–83. Routledge, Howe (2010)
6. Baldwin, C.: Narrative(,) citizenship and dementia: the personal and the political. J. Aging Stud. **22**(3), 222–228 (2008). https://doi.org/10.1016/j.jaging.2007.04.002
7. Dran, D.S.: A new look at episodes of mistaken identity: opportunities for preserving personhood. J. Appl. Gerontol. **27**(5), 641–647 (2008)
8. Hubbard, G., Cook, A., Tester, S., Downs, M.: Beyond words: older people with dementia using and interpreting nonverbal behaviour. J. Aging Stud. **16**, 155–167 (2002)
9. Hydén, L.-C.: Storytelling in dementia: embodiment as a resource. Dementia **12**(3), 359–367 (2013). https://doi.org/10.1177/1471301213476290
10. Kontos, P.C.: Ethnographic reflections on selfhood, embodiment and alzheimer's disease. Ageing & Soc. **24**(6), 829–849 (2004)
11. Kessler, L.: Dancing with rose: finding life in the land of alzheimer's. Viking, New York (2007)
12. Novy, C.: Themes of care giving and receiving in the interconnecting stories of a mother living with dementia and her adult son: a single-case life story study. J. Aging Stud. **46**, 32–36 (2018). https://doi.org/10.1016/j.jaging.2018.06.001
13. Novy, C.: Life stories and their performance in dementia care. Arts Psychother. **57**, 95–101 (2018). https://doi.org/10.1016/j.aip.2017.12.003
14. Dewing, J.: From ritual to relationship: a person-centred approach to consent in qualitative research with older people who have dementia. Dementia **1** (2), 157–171 (2002)
15. Bartlett, R., O'Connor, D.: Broadening the dementia debate: towards social citizenship. The Policy Press, Bristol, UK (2010)
16. Kitwood, T.: Dementia reconsidered: the person comes first. Open University Press, Buckingham (1997)
17. Isaacs, E.A., Clark, H.H.: References in conversation between experts and novices. J. Exp. Psychol. **116** (1), 26–37 (1987)
18. Schweitzer, P., Bruce, E.: Remembering yesterday, caring today. Reminiscence in dementia care. A guide to good practice, p. 30. Jessica Kingsley, London (2008)
19. Pelias, R.J.: Performative inquiry: embodiment and its challenges. In: Knowles, J.G., Cole, A.L. (eds.) Handbook of the arts in qualitative research: perspectives, methodologies, examples, and issues, pp. 186–194. Sage Publishing, Thousand Oaks, CA (2008)
20. Baldwin, C., Greason, M.: Micro-citizenship, dementia and long-term care. Dementia **15**(3), 289–303 (2016). https://doi.org/10.1177/1471301216638762
21. Brannelly, T.: Citizenship and people living with dementia: a case for the ethics of care. Dementia **15** (3), 304–314 (2016). https://doi.org/10.1177/1471301216639463
22. Tronto, C.: Caring democracy: markets, equality, and justice. New York University Press, New York (2013)
23. Hughes, T., Castro Romero, M.: A processural consent methodology with people diagnosed with dementia. Qual. Ageing Older Adults **16**(4), 222–234 (2015)
24. Luff, R., Laybourne, A., Ferreira, Z., Meyer, J.: A guide to research with care homes. Qual. Ageing Older Adults **16**(4), 186–194 (2015). http://dx.doi.org/10.1108/QAOA-06-2015-0027
25. Armstrong, P.: Skills for care. In: Amstrong, P., Braedley, S. (eds.) Troubling care. critical perspectives on research and practices, pp. 101–102. Canadian Scholar's Press Inc., Toronto (2013)
26. Day, S.: The implications of conceptualizing care. In Amstrong, P., Braedley, S. (eds.) Troubling care: critical perspectives on research and practices, pp. 21–31. Canadian Scholar's Press Inc., Toronto (2013)
27. Tronto, J.C.: An ethic of care. Generations **22**(3), 15–20 (1998)
28. Brown Wilson, C., Davies, S.: Developing relationships in long term care environments: the contribution of staff. J. Clin. Nurs. **18**(12), 1746–1755 (2009)
29. Hubbard, G., Downs, M.G., Tester, S.: Including older people with dementia in research: challenges and strategies. Aging & Ment. Health **7**(5), 351–362 (2003). https://doi.org/10.1080/1360786031000150685
30. Hydén, L-C.: Entangled narratives: collaborative storytelling and the re-imagining of dementia. Oxford University Press, Oxford (2017)
31. Downs, M., Collins, L.: Person-centred communication in dementia care. Nurs. Stand. **30**(11), 37–41 (2015)
32. Ward, R., Campbell, S.: Mixing methods to explore appearance in dementia care. Dementia **12**(3), 337–347 (2013)
33. Buse, C., Twigg, J.: Materialising memories: exploring the stories of people with dementia through dress. Ageing & Soc. **26**, 1115–1135 (2016)
34. Hung, L., Phinney, A., Chaudhury, H., Rodney, P., Tabamo, J., Bohl, D.: "Little things matter!" exploring the perspectives of patients with dementia about the hospital environment. Int. J. Older People Nurs. **12**(e12153), 1–11 (2017). https://doi.org/10.1111/opn.12153
35. Holmes, D., Murray, S.J., Perron, A., Rail, G.: Deconstructing the evidence-based discourse in health sciences: truth, power and fascism. Int. J. Evid.-Based Healthc. **4**(3), 180–186 (2006)
36. Hogan, S., Pink, S.: Visualising interior worlds: interdisciplinary routes to knowing. In: Pink, S. (ed.) Advances in Visual Methodology. Advances in Visual Methodology, pp. 230–247. Sage Publications, London (2013)

Estonian Museums and People with Dementia: Experiences and Challenges

Anna-Kristina Rätsep and Ester Bardone

Abstract

Museums are part of a new movement by different cultural organisations to become more accessible to diverse audience groups including people with dementia. At the same time, very few studies have examined the potential role that museums may play in improving the quality of life for persons with dementia and their caregivers. In Estonia the number of people with dementia is rising and currently there are over 22,000 people diagnosed with dementia. However, until now, the issue has not been addressed in Estonian museums. This ethnographic study, will be the first one to raise the awareness of this topic in Estonia. The paper discusses the growing role of design in dementia care and its power to enhance the wellbeing of people living with dementia and their caregivers. In this study ethnographic research methods are applied—participant observation experience and eleven expert interviews in total, eight with Estonian museum workers and three with occupational therapists working with PWD. The informants were chosen from two cities—Tallinn and Tartu, which have the largest and most visited museums in Estonia. The aim of this study is to include and compare mixed selection of state owned museums carrying the largest art and ethnographic-historical collections [Estonian National Museum], art collections [Art Museum of Estonia and Tartu Art Museum] and collection of historical artefacts [Estonian History Museum]. Three main themes arise from the fieldwork which are being discussed: (a) museum staff experience with people with dementia (b) museum staff training possibilities and potential partners and (c) museum´s challenges; cultural policy and financial resources.

Keywords

Museums · Estonia · Dementia · Wellbeing

1 Introduction

Museums have not been part of the public health realm until recently, as their focus in last decades has been on education and inclusion rather than issues of health and wellbeing [1]. However, throughout these years museological debate has increasingly focused on whether, in addition to heritage institutions, museums should be social service institutions performing “a bolder role as agents of wellbeing and vehicles of social

A.-K. Rätsep · E. Bardone (✉)
University of Tartu, Tartu, Estonia
e-mail: ester.bardone@ut.ee

R. Brankaert et al. (eds.), *Dementia Lab 2021: Supporting Ability Through Design*, Design For Inclusion 2, https://doi.org/10.1007/978-3-030-70293-9_10

change" [2, pp. 3–4]. Furthermore, it has been suggested that as "socially purposeful" institutions museums may "challenge public perceptions of disability and difference" [3]. These claims have been motivated both by policies, academic research as well as museums' own reflections about the functions they should have in modern society and can be framed in a broader discussion about cultural democracy, cultural participation and civic engagement.

The UK is a good example of how, due to national cultural and health policies, museums have gained a noticeable position in public health and wellbeing and throughout the country a considerable diversity of respective museum programmes and projects are provided in collaboration with organizations from the health and social care sector [4]. Several museums in the UK, US and Australia have demonstrated their capacity to contribute to social inclusion by involving elderly in telling different stories about old age and co-creating programmes for people who suffer cognitive decline, including people with dementia [hereafter PWD] [5].

Such developments can be related to increasing public recognition of preventive and non-pharmacological healthcare and the therapeutic role that arts, culture and creative experiences may have in this context. Furthermore, currently the effectiveness of the pharmacological treatment of dementia is moderate at best and therefore non-pharmacological interventions play an especially important role in the care and general wellbeing of PWD. Recent research has shown that reminiscence therapy, physical activity, arts therapy, psychosocial and education activities in groups etc. have raised the quality of life[1] for dementia sufferers and enable treating behavioural and psychological symptoms with no adverse effects [6].

Museums are "nearly always non-stigmatising settings" [7, p. 66] that differ from clinical and social care setting and may provide multisensory engagement, such as tactile stimulation which encourages reminiscence and memory that can affect emotions, ideas of self-esteem and general sense of well-being of PWD [7, 8]. For instance, the material, physical and intrinsic properties of museum objects "trigger a variety of emotional and sensory responses, cognitive associations, memories and projections". Museum object handling in small group settings has proved to have a positive impact on the subjective well-being of people with different types and stages of dementia[2] [9, p. 795]. Previous research conducted has demonstrated that attending museum programmes designed for PWD not only supports the social wellbeing of those who suffer from dementia but can be satisfying also for caregivers and the staff involved [i.e. 10–14]. Similarly, museum programmes for PWD have demonstrated that viewing works of art and making art by yourself has fostered diverse aspects of wellbeing of PWD who described elevation in their self-esteem and their carers reported elevated mood and increase in social support [7, 13].

As stated in Dementia in Europe Yearbook [14], Eastern European countries lack National Dementia Strategies. According to the World Health Organisation currently in Estonia there is very low level of a public awareness regarding dementia and available guidance for health or social care staff to identify and reduce carer stress. Furthermore, research on dementia is very modest, rarely involving PWD in the research process [15]. There is also a lack of national statistics about PWD as the number of them is not regularly monitored [according to the available data there are over 22 000 people diagnosed with dementia] and a considerable number of cases remain undiagnosed [16]. At the same time, similarly to many European countries, Estonia is a fastly aging society and it is estimated that by the year 2035 the number of elderly people (>65) will comprise 25% of the population [17]. Thus, raising public awareness

[1]Quality of life (QOL) is an overarching term for the quality of the various domains in life. It is a standard level that consists of the expectations of an individual or society for a good life. These expectations are guided by the values, goals and socio-cultural context in which an individual lives (WHO 1998).

[2]For those at milder levels of impairment such interventions have shown the greatest wellbeing improvement.

of dementia in Estonia is of vital importance. Currently an activist agency NGO Life With Dementia [established in 2016] has played a key role in organising public campaigns and support groups for PWD and their carers.[3] A recent study examining museums' visitors and non-visitors expectations in Estonia revealed that that health related services could belong to the future of Estonian museums as about 60% of respondents agreed that they would potentially visit museums more often if they could participate in activities and events that would help to improve their health and wellbeing [18].

Although Estonian society has just started raising public awareness about dementia and dealing with non-pharmacological interventions in dementia care, we believe that there is a need to examine how Estonian museums could contribute to improving the wellbeing of PWD and their caregivers. This study is part of an ongoing doctoral research project, delivered by the first author Anna-Kristina Rätsep [hereafter AKR] and assisted by the co-author, Ester Bardone [hereafter EB], focusing on PWD and their caregivers as a potential audience group for Estonian museums and more specifically dealing with issues like cultural accessibility, co-design of museum programmes and wellbeing of these people. The paper presents the preliminary results and focuses on two major research questions: (1) What are Estonian museums professionals' experiences with PWD, [2] What are their current training possibilities and potential partners for being able to work with PWD and a potential audience group? (2) What are the challenges museums face in aiming to develop programmes for PWD and their caregivers considering Estonian cultural policies and museums' financial resources?

[3]According to an interview with an occupational therapist Hanna-Stiina many caregivers, as well as friends and family, of persons with Alzheimer's disease or another type of dementia find support groups to be a wonderful source of information, encouragement, and inspiration. In addition, many persons who have recently been diagnosed with dementia or are in the early stages of the disease, find support groups helpful.

2 Materials and Methods

In order to examine Estonian museum staff's current experience of and dispositions towards engaging with PWD, AKR conducted interviews with museum professionals (9 interviews) and occupational therapists working with PWD (3 interviews) [see Table 1]. We selected informants from two major cities in Estonia (Tallinn and Tartu), which have the largest number of habitants, including also the largest and most visited museums [19]. The aim was to include and compare a mixed selection of state-owned museums carrying the largest art and ethnographic collections [Estonian National Museum], art collections [Art Museum of Estonia and Tartu Art Museum] and collection of historical artefacts [Estonian History Museum]. As PWD is a potential audience group in museums engaged mainly by education specialists, the inclusion criteria of our interviewees required an occupation in the field of Museum Education which could enable us to reflect upon potential activities and themes to be delivered for PWD. Also, two Directors of Developments were included as they carry a broader perspective of the functioning of a museum including the museum′s financial aspects. Furthermore, occupational therapists were involved in this study through the NGO Life with Dementia which has a membership among diverse health professionals. The inclusion of occupational therapists was based on the fact that in Estonia there are more occupational therapists involved working with PWD than art therapists. Occupational therapists are organised in special social institutions, hospices and hospitals, providing activities and care for PWD. Initially, all the interviews were planned as face-to-face interviews, however, due to physical and travelling restrictions implemented by Covid19, the majority of interviews were conducted via Skype in between February 2020 until April 2020. All the informants gave their consent for the publication of their name and occupation.

Additionally, AKR conducted participant observation attending special events in the Kumu Museum "Re-encounters" [Est. *Taaskohtumised*]

Table 1 Information about research participants of this study

Interviewee	Position/Occupation
Anti Lillak	Museum Educator in Estonian National Museum [ENM]
Kaari Siemer	Head of Museum Education in Estonian National Museum [ENM]
Inge Laurik-Teder	Head of Estonian Museum Association Board, Director of Development in Estonian History Museum [EHM]
Liisi Selg	Head of Museum Education in Estonian History Museum [EHM]
Kersti Tiik	Director of Development in Art Museum of Estonia, including all its five branches [AME]
Sandra Lääne-Velt	Head of Education in Kumu museum, one of the five branches of The Art Museum of Estonia [AME]
Hanna-Liis Kont	Former employee of the Tartu Art Museum [TAM]
Kristel Sibul	Museum Educator in Tartu Art Museum [TAM]
Hanna-Stiina Heinmets	Occupational therapist in The Estonian Evangelical Lutheran Church Diaconial Hospital [EELC Diaconical Hospital]
Talvi Toom	Occupational therapist in Tallinn Health Care College
Liis Niilo	Occupational therapist in Villa Benita Hospice

programme as well as in "Memory Café" in Saue Daily Center in order to learn what type of activities are provided for PWD, how is the physical space accommodated and how the staff engages with PWD and their carers. Observations helped to make sense of how PWD express themselves in a museum setting as well as in daily centers, e.g. what type of memories and stories PWD shared about museum artefacts or historical memorabilia. AKR participated in some of the events assisting museum workers in preparing the education space and provided help for delivering the crafting activities in the educational space.

Fieldnotes and interview data was analysed by using thematic analysis, assisted by Express Scribe qualitative software. Thematic analysis resulted in the following overarching main themes: professional background of interviewees; museum´s main audience groups, adult education, and elderly people; new audience groups in the future; museums workers' personal experience with people with dementia and their training for dealing with this group, museum´s resources for developing programmes for PWD [human, physical, financial resources], potential partners for developing programs for PWD, potential themes and activities for PWD, museum´s benefits for PWD and vice versa. For contextualising the issues present in interview data, we analysed various official documents that enabled us to see the role of cultural policies in Estonia in shaping how PWD and their caregivers as a museum audience group is acknowledged. As follows we will focus on some of the themes more closely.

3 Estonian Museums and Aging Adults

Museum specialists reported the importance and the role of engaging with "aging adults" whom they see as frequent visitors at museum exhibitions as well as participants in diverse adult programmes, having the resource of "curiosity" and "free time". Even though, elderly have the resources to be engaged actively in museums, Estonian museum audience research conducted in 2018, showed that the main visitors currently in Estonian museums are instead younger people in the age between 15–19 and middle aged in between 30–44, whereas seniors in the age 75 + can are mainly “non-visitors” [19%] who have not visited any museum within the past 12 months. The main reasons why the respondents have not been in the museum are the following: firstly, the lack of time [41%], secondly, health reasons [15%] and thirdly, inconvenient location and problems of accessibility [11%] [18]. At the same time, Estonian museums are starting to value the importance of elderly and some of them are now also developing special programs for them. For instance, Kumu & EHM have initiated special senior programmes, whereas ENM does not see the need to establish specific activities for seniors and consider them as part of adult audiences. Furthermore, museums value elderly people’s stories as part of oral history that needs to be preserved. The EHM considers collaboration with elderly people a gratifying as well as rewarding experience and engages them in exhibition development processes by collecting their stories and artefacts [Inge-Laurik Teder].

4 Museum Staff Experience with People with Dementia

PWD are a specific part of the senior audience that needs specially designed and delivered programmes which accommodate to their needs. Our results demonstrate that in Estonia, this work is still in its infancy and requires ongoing attention. Several Estonian museum workers admitted having limited knowledge regarding the nature of dementia and except one interviewee [Sandra Lääne-Velt, Kumu], others have not yet any experience in developing programs for PWD and their caregivers. Furthermore, informants also had limited personal experience with PWD as only two interviewees had a personal contact with an elderly relative diagnosed with dementia, some others reported about memory issues among their elderly family members who were not diagnosed with dementia. Museum staff expressed the need to be informed and educated about PWD and their needs, specifically in the museum context. In effect, one of us [AKR] shared information about the symptoms and types of dementia before and during the interviews which enabled research participants to think about the possibilities and limitations of museums to engage people with dementia.

Even though museum workers did not have much personal or professional experience with PWD, they are all interested in starting to collaborate with this new audience group valuing its importance to their museum. Above all, staff members see it as a reciprocal relationship—museums carry an important role for PWD and vice versa, PWD are important for the museum’s mission to be a social institution in the society, serving the needs of marginalised groups. For instance, elderly people, especially PWD, have a high risk of being socially isolated [20]. For this reason, museums may carry a social value for such people providing opportunities for social interactions, for expressing themselves and feeling connected to others [Kristel Sibul, TAM]. According to Sandra Lääne-Velt [Kumu], museums in Estonia could become “meeting places” for elderly and PWD together with their carers where they feel “welcomed” and “included”, artworks especially may carry therapeutic effect and “evoke emotions and memories, make people converse and recollect”. Sandra suggests that museums become important environments that give the person back certain aspects of their selfhood, such as the feeling of “independence” and “control” [Kumu].

Additionally, museum staff emphasised that museums can be supportive places for PWD in order to socialise with each other and also, where reminiscence work is carried out by interpreting objects to enable elderly make meaning from them by sharing life experiences, memories and stories from their past. Typically, PWD are more able to recall things from many years ago than recent memories, so reminiscence draws on this strength. One of our informants also highlighted the importance of the museum as a place for reminiscence:

> I think that the reason why museums are relevant for people with dementia is the fact that museums are full of memories. Museums are full of memories and different time periods. Kumu has four different exhibition floors and we find here

centuries of Estonian art history. In fact, there are a lot of these things which we can use for revitalizing memories and also, to create new memories. [Sandra Lääne-Velt, Kumu]

A museum educator Anti Lillak[ENM] believes that doing reminiscence work in the museum carries an important role also for supporting caregivers of PWD who have a high burden of care concerning everyday practicalities and no energy to provide memory stimulating activities. However, so far only one specific reminiscence type of museum programme has been designed in Estonia for PWD and their caregivers. In 2017, Kumu had initiated in collaboration with NGO Life with Dementia a museum program *Re-encounters* that has been delivered from 2018–2019 [Kumu, 2020; Lääne-Velt, 2020]. This program was designed by the Department of Education and offered under the category of "public program". According to Sandra Lääne-Velt, the aim of the programme was to provide the participants socially relevant and emotionally engaging activities by applying art therapy and storytelling methods. The "Re-encounters" program is offered twice a month on Saturday mornings, when ca 15 PWDs come to the museum with their carers. The session comprises a 45-min museum tour followed by a 30 min hands-on crafting session in the museum education space. AKR observed non-verbal responses and reactions of PWD to museum workers and their carers; and informal discussions and interactions in between PWD and their carers. Observations showed that these sessions had a very strong emotional impact on participants, as stories told during these sessions were personal and often revealed the participants´ personalities, interests and life-histories. Participants showed positive emotions such as smiling and hugging the staff members at the end of the event. At the same time, not all PWD were open to speak and share their stories with others around and hence, needed the encouragement of the staff. The sessions also fostered physical activity in participants enabling them to stimulate their hand-motorics and cognitive skills. Sandra Lääne-Velt also engaged PWDs and their caregivers in the co-curation process of the program by giving them the chance to choose which paintings from Kumu collections they may wish to see during the next session. For this, she created a curated set of reproductions to be made a selection amongst. Sandra claimed the co-design as essential part of the programme preparation process—this way PWDs had a chance to "feel independent", to be able to "make a choice themselves" and "to establish an immediate personal connection with a particular painting".

5 Museum Staff's Training Possibilities and Potential Partners

Our fieldwork gave us insights that the first and foremost resource required for the new museum program for PWD is the knowledge regarding the needs of the audience group. While interviewed occupational therapists mentioned the relevant knowledge of communication techniques with PWD, our informants in museums admitted not having any previous experience working with PWD and their carers. As limited knowledge exists among museums regarding this audience group, participants confirmed the need for training museum staff which would help them to make sense of their new audience group and provide understanding and specific methods on how to work with PWD and their caregivers in museum space. Also, occupational therapists confirmed the importance of training before starting a program for PWD. During our fieldwork, the question was raised: What are the best ways to provide this type of training for Estonian museum workers? According to our informants, there exist extensive and varied methods for professional training. Above all, museum practitioners mentioned short-term training opportunities such as one day workshops or information days.

Compared to museum specialists, occupational therapists demonstrated a more interdisciplinary understanding of potential partners, emphasising on the importance of collaborating with diverse social care partners for providing museum training. Liis Niilo [Villa Benita

Hospice] highlighted the need for a more integrated approach and the importance of working in collaboration with different caregiving professionals [e.g. psychologists, occupational therapists and social workers] to inform museum workers about the characteristics of PWD. Furthermore, it is very important to include in the training sessions people who have already participated in organized museum events for PWD, as sharing of their personal experience with dementia would allow the museum workers to understand the importance and benefits of their future museum programs. Also, occupational therapists suggested that museum workers could participate in events outside museum spaces provided for instance by the local daily care centers in order to gain new experience and courage in how to communicate with PWD. In fact, one of the museum workers, Kaari Siemer [ENM] claimed that the role of Estonian museums as partners for healthcare institutions, such as mental health hospitals, has risen in adult education. In 2017, a specific project was developed by ENM in order to aid the wellbeing of people with mental disorders in the Psychiatry Clinic in Tartu. This project occurred to be very successful bringing new knowledge and experience to museum workers and also, providing an exciting learning and social environment for people with mental issues. Similarly, to programs for PWD, the need for a pre-training session was brought out by Kaari Siemer [ENM], which was at that time organised as an off-site visit to the Psychiatry Hospital, during which museum workers were able to familiarise themselves with the characteristics and needs of the patients. At the same time, collaboration is also important on a cultural institutional level. Another possibility to provide training for museum staff is to collaborate with the National Heritage Board of Estonia, which emphasises on the value of cultural heritage, including environments of cultural value such as museums. In fact, the National Heritage Board has become, since 2019, a central institution advising and providing training for museum workers [21] and can therefore become also one potential partner for developing new programs in museums for PWD.

Our fieldwork also raised another question: What type of knowledge should be provided for museum staff during these training sessions? Occupational therapists have extensive experiences regarding the daily needs of PWD and their caregivers and according to them, museum staff should take into consideration several following aspects of their museum space and staff while developing programs for PWD. Above all, Hanna-Stiina Heinmets [EELC Diaconical Hospital] stressed that museums must ensure that all their departments are welcoming and accessible, as it is the responsibility of the whole staff to be aware of the specific needs and characteristics of PWD. Museum staff, including exhibition guards, need to be well informed about the characteristics of dementia and how to behave with PWD during the sessions. Above all, how visitors are greeted and treated by museum staff during their visit is of vital importance as they leave an emotional effect on PWD. In fact, it was interestingly brought out by HS that often the content of the program may be forgotten by PWD, but the positive feeling towards the program deliverer may have a long-lasting effect. Therefore, good educators, who deliver the program are essential for a successful program. In fact, educational specialists need to develop certain communicative skills while delivering museum programs for PWD. First and foremost, introduction is the most important part of the session as it enables to create a relaxed and comfortable mood in the participants. At the beginning of the sessions the educator should introduce herself, offer participants some coffee and provide logistical information of the location of toilets in case they may need it urgently. Furthermore, it should be taken into consideration that old people may be slower in thinking and they sometimes have complications with speech. Hence, the deliverer needs to acquire patience, adapt to the walking and speech tempo of the elderly. Also, it is important to consider that the processing of incoming information for a PWD will become impaired in time, in that the individual may find it increasingly difficult to attend to more than one stimulus at a time, such as taking part in conversations with several persons and shifting of the focus of attention from one topic to another. This idea was brought out by Liis Niilo [Villa Benita Hospice] who believed that it may be hard for PWD to perform two or more activities simultaneously, hence, recommending the guide to make a stop while they introduce certain objects in the gallery. In this way, she believes that PWD can concentrate better on the guide´s verbal speech:

> He / she cannot walk and speak or walk and be engaged in a conversation simultaneously. From a certain level, it is difficult to conduct two activities at the same time. Walking and observing for example are two different activities. Or walking and speaking. When the moment arrives where one must listen and think along, they will often stop and continue then on speaking. This is an interesting thing that one should pay attention to.

6 Museums' Challenges: Cultural Policy and Financial Resources

The Estonian Museum Act relies on the ICOM museum definition stressing, among other aspects, that museums should respond to the needs of disabled people [19]. Furthermore, in the development document of Estonian cultural policy until the year 2020, a goal was set [for 2014–2020] to create better opportunities for disabled people to participate in culture, considering the needs of people with disabilities and in supporting diverse initiatives and programs in cultural institutions aimed at people with physical, mental or other disabilities [22]. This promise raises the question: To which extent is this goal put into practice engaging disabled people in Estonian museums? While reading the official documents of the Ministry of Culture it appeared that the main developments regarding Estonian museums´ accessibility to disabled groups have been in their physical space. Often these spatial changes have been initiated by new museum exhibitions which have brought along the need to re-think their physical spaces and better accessibility to disabled groups [ibid.]. Furthermore, steps have been made towards engaging disabled groups within museum programmes. Our fieldwork materials suggest that audience groups participating in educational activities are becoming more diversified and museums are now engaging different vulnerable groups. Above all, museum workers regarded their museums as "informal learning environments", conveying and thus, supporting the educational process of children. This is also supported by official institutional development documents of studied museums which highlight the importance of museum educators in curating and delivering educational activities [SA Eesti Kunstimuuseumi Arengukava 2017–2020[4]; ERM 2017–2020,[5] Tartu Kunstimuuseum. 2020]. However, Kersti Tiik [EAM] highlighted that even though museums are valuable places which support schools in achieving their learning outcomes, this type of activity is not profitable compared to other museum activities which bring much more financial benefits. For instance, according to Kersti Tiik, EAM´s exhibition development raises their private revenue 63%, museum shop and events management 30%, whereas educational activities form only 7% of their own revenue.

Interviewed museum staff claimed that museums are supported by the Estonian state directly from the state budget and local government budgets. However, the future budgetary planning can be decided by the museum for fulfilling the main tasks of the museum [23]. Our informants believed various funding shcemes exist which could aid the implementation of a museum programs for PWD and their caregivers. On the one hand, they found that it should be the role of the Estonian state and governmental institutions [Ministry of Culture or Ministry of Social Welfare] to provide funding for museums to implement programs for people who have physical and cognitive disabilities. On the other hand, some informants suggested that the museums themselves should set and target goals on how they wish to finance these new programs for PWD. In fact, PWD are seen by our informants as a socially valuable audience group and in accordance with the museums´ developmental

[4]https://kunstimuuseum.ekm.ee/wp-content/uploads/2019/12/Arengukava.pdf.

[5]https://www.erm.ee/et/content/arengukava-2017-2020.

plan and priorities to include children and adult groups with special needs. Therefore, it is believed that support for this program may be planned in the budget of the Department of Education. For example, until now, Kumu followed this admission logic partly by providing free entrance to the "Re-encounters" program for PWD who had a disability card. On the other hand, a reduced admission ticket was offered for their caregivers. Sandra Lääne-Velt [Kumu] believes that when external financing opportunities are not available then funding for programs for PWD can be provided from the Art Museum budget itself. Lääne-Velt considers possible to implement a dementia program with less economic profit:

> Our educational department delivers different programs and there needs to exist a balance - if we spend a lot of money into one action, we need to earn more money from other sources. Definitely "Re-encounters" is not a program which would bring in a lot of money. Unfortunately, this program has not received much finance from outside and for this, we need to be more creative to deliver it ourselves with our own staff and using materials which we already have previously.

Furthermore, it was noted that it may become especially difficult for smaller museums with less human and financial resources to implement new programs for vulnerable groups. In order to make programs for PWD sustainable in the long-term, many informants mentioned the importance of extra financial resources such as special government grants. Kaari Siemer [ENM] believed that the finance for programs for PWD may come in the future for their museum from different project funding sourced either by the Cultural Ministry or Ministry of Social Affairs. At the same time, specific concerns were also mentioned regarding grant sources for these types of programs.

In closing, programs for PWD add to existing museum costs and need therefore special funding sources from the state. It is therefore of relevant importance to develop in the next following years these additional funding sources in order to support the implementation of PWD programs in Estonian museum.

References

1. Chatterjee, H., Noble, G.: Museums, health and well-being. Routledge, London (2016)
2. Silverman, L.H.: The social work of museums. Routledge, London (2009)
3. Dodd, J.: The socially purposeful museum. Mus. Logica Brun. **4**(2), 28–32 (2015)
4. Morse, N.: The social role of the museums: from social inclusion to health and wellbeing. In: O'Neill, M., Hooper, G. (eds.) Connecting museums, pp. 48–65. London and New York, Routledge (2020)
5. Ward, E.: Transforming health, museums and the civic imagination. In: O'Neill, M., Hooper, G. (eds.) Connecting museums, pp. 125–137 (2020). London and New York, Routledge (2012)
6. Beerens, H., Sutcliffe, C. Renom-Guiteras, A. et al.: Quality of life and quality of care for people with dementia receiving long-term institutional care or professional home care: the european right time place care study. J. Am. Med. Dir. Assoc. (2014) Available online at: http://www.sciencedirect.com/science/article/pii/S152586101300546X
7. Roberts, S., Camic, P.M., & Springham. N.: New roles for art galleries: art-viewing as a community intervention for family carers of people with mental health problems. Arts & Health: Int. J. Res., Policy Pract. **3**, 146–159 (2011)
8. Eekelaar, C., Camic, P.M., Springham, N.: Art galleries, episodic memory and verbal fluency in dementia: an exploratory study. Psychol. Aesthet., Creat., Arts. **6**(3), 262–272 (2012)
9. Camic, P.M., Hulbert, S., Kimmel, J.: Museum object handling: a health-promoting community-based activity for dementia care. J. Health Psychol. **24**(6), 787–798 (2019)
10. Roe, B., McCormick, S., Lucas, T., Gallagher, W., Winn, A., Elkin, S.: Coffee, cake & culture: evaluation of an art for health programme for older people in the community with dementia. Dementia. **15**(4), 539–559 (2016)
11. Camic, P.M., Baker, E. Tischler, V.: Theorising how art gallery interventions impact people with dementia and their caregivers. Gerontol., 1–10 (2015)
12. Flatt, J.D. et al.: Subjective experiences of an art museum engagement activity for persons with early alzheimer's disease and their family caregivers. Am. J. Alzheimer's Dis. Other Dement. **30**(4), 380–389 (2015)
13. Smiraglia, C.: Qualities of the participant experience in an object-based museum outreach program to retirement communities. Educ. Gerontology. **41**(3), 238–248 (2015)
14. Camic, P.M., Tischler, V., Pearman, C.H.: Viewing and making art together: a multi-session art gallery based intervention for people with dementia and their carers. Aging & Ment. Health **18**(2), 161–168 (2014)

15. MacPherson, S., Bird, M., Anderson, K., Davis, T., Blair. A.: An art gallery access programme for people with dementia: 'you do it for the moment'. Aging & Ment. Health **13**, 744–752 (2009)
16. Alzheimer Europe.: Dementia in Europe yearbook. comparison of national dementia strategies in Europe (2018). https://www.alzheimer-europe.org/Policy/Country-comparisons/2018-Comparsion-of-National-Dementia-Strategies, last accessed 10 July 2020
17. World Health Organization.: Dementia (2020). Available online at: https://www.who.int/news-room/fact-sheets/detail/dementia#:~:text=Dementia%20is%20a%20syndrome%20in,million%20new%20cases%20every%20year, last accessed 10 July 2020
18. Saks, K.: Inimese elukaar. Saks, K (Editors Abbr). Gerontoloogia (11–18). TÜ Kirjastus (2016)
19. Kultuur.: Aruanne "Kultuuripoliitika põhialused aastani 2020" 2019 aasta täitmise kohta (2013). Available online at: https://www.kul.ee/sites/kulminn/files/200115_lisa1_kultuur2020_aruanne2019.pdf, last accessed 10 July 2020
20. Kantar Emor.: Muuseumide ja raamatukogude külastajate ja mittekülastajate uuring. Uuringu lõpparuanne (2018). Available online at: https://www.kul.ee/sites/kulminn/files/muuseumide_ja_raamatukogude_kulastajate_ja_mittekulastajate_uuringu_lopparuanne_0.pdf, last accessed 10 July 2020
21. Republic of Estonia.: Government Office (2019). Available online at: https://www.riigikantselei.ee/sites/default/files/riigikantselei/strateegiaburoo/Eesti2035/AV_paberid/av_edukalt_kohaneda_muutustega_rahvastikus.pdf, last accessed 30 October 2020
22. Börsch-Supan, A.: Ageing in Europe—supporting policies for an inclusive society (2015). https://www.degruyter.com/view/title/514422, last accessed 10 July 2020
23. Museums Act.: (2013). Available online at: https://www.riigiteataja.ee/en/eli/510012014002/consolide, last accessed 10 July 2020

Co-creating Design Opportunities for Social Technology in the Context of Dementia

Agnieszka Barbara Kolasinska, Myrte Thoolen, Sebastiaan Peek, Yuan Lu, and Rens Brankaert

Abstract

Maintaining social participation in late life is an important component of general well-being. For people with dementia, participating in social interactions may become increasingly challenging with the progression of the condition. Social technologies can potentially support and empower them in this setting, yet there is a need to get a comprehensive understanding of how their social participation is hindered and facilitated through technology. This paper presents the outcomes of four co-creation sessions conducted together with four couples. We explored their experienced social challenges and identified opportunities for design to improve their social participation. These focus on compensating for internal and external barriers through design, enhancing agency through the social and physical environment, and learning from previous experiences with existing social technologies. Furthermore, we provide recommendations for future research in the context of social technologies for people with dementia. We argue that design-research should focus on embracing the importance of a broader social context, sharing meaningful activities, and incorporating social contact into the everyday routines. The significance of these outcomes indicates the relevance of well-designed novel social technologies to enrich people's with dementia experience in the social context.

Keywords

Dementia · Co-creation · Design · Experience · Social technology

The first two authors contributed equally to the paper.

A. B. Kolasinska · M. Thoolen (✉) · S. Peek · Y. Lu · R. Brankaert
Department of Industrial Design, Eindhoven University of Technology, Eindhoven, The Netherlands
e-mail: m.e.thoolen@tue.nl

A. B. Kolasinska
e-mail: agnieszkabarbara.kolasinska@phd.unipd.it

A. B. Kolasinska
Department of General Psychology, University of Padua, Padua, Italy

S. Peek
School of Social and Behavioral Sciences, Tilburg University, Tilburg, The Netherlands

R. Brankaert
School for Allied Health Professions, Fontys University of Applied Sciences, Eindhoven, The Netherlands

R. Brankaert et al. (eds.), *Dementia Lab 2021: Supporting Ability Through Design*, Design For Inclusion 2, https://doi.org/10.1007/978-3-030-70293-9_11

1 Introduction

1.1 Social Interaction Within the Context of Dementia

Dementia, particularly Alzheimer's type of pathology, affects an increasing number of people all over the world. It is a syndrome that causes cognitive impairment characterized by progressive deterioration in abilities and behavior, affecting activities of daily living, social resilience [1] and altering former levels of functioning [2]. The challenges associated with dementia are known to negatively influence social interactions of people affected by the condition making it difficult to maintain social relationships and the sense of individual contribution to society [3]. Research shows that people with dementia tend to withdraw from social contacts, especially concerning larger groups of people [4]. Similarly, previous studies reveal that one of the most frequently experienced unmet needs of people with dementia is having other people's company [5] [6]. Social context, defined not only by the size of the social network or frequency of social interactions, but most importantly by the personal experience of connection and meaning in social interactions, is crucial for promoting well-being and maintaining and (re)constructing our identities [7] [8] [9] [10]. Kitwood and colleagues (1992) argue that personhood is created through social participation and interactions with others [9]. It can be understood as the attributes possessed by humans that make them a person [11]. It is of particular importance within the context of person-centered care. According to Kitwood [12], a person-centered approach should emphasize communication and relationships, as opposed to medical and behavioral perspectives. Person-centered practice encourages care professionals to put the person, not the evidence-based interventions, first [13]. One of the crucial assumptions of the person-centered approach is seeing the person with dementia as an agent capable of taking control of their own life in a meaningful way [9]. Agency is not simply an attribute of humans. It is being generated through configurations of people and their interactions with each other and with their environment [14] [15] [16] [17]. Therefore, we argue that the agency of people with dementia can be supported and empowered by their surrounding social and technological environment.

1.2 Design and HCI in Dementia Care

In line with the emerging person-centered paradigm in dementia care, a shift has been taking place in the field of HCI and dementia with regard to a change in perspective [18], from supporting people who are experiencing cognitive and physical impairments to addressing a persons' experience [19] [20] [21] [22]. Given the changing nature and the complexity of the condition, research in HCI emphasizes the need to understand and support the maintenance of 'personhood' of people with dementia [23] [8] [24] [25], by acknowledging that they are still persons that can experience life and relationships in full effect. Based on this person-centeredness, many HCI researchers have emphasized the value of inclusive design practices [24] [19] [26] [27], enabling them to emphatically engage with people with dementia [8], to best cater to their needs and to anticipate their abilities by design [28]. The perception and reality of the individual is hereby highly valued, and is being investigated through experience-centered design (ECD) [29]. It is increasingly being applied in HCI to discover how people with dementia experience their social world [21] [25] [27], and how design and technology can enrich the interactions in this social environment [30] [22] [17].

Building a connection in a participatory way is fundamental for gaining a deeper understanding in the experience of people with dementia and their perspective on technology [23] [24] [31] [32]. However, traditional participatory practice is too demanding [33] [34], since it relies on the use of verbal expression, abstract thinking [18], and demands resilience and initiative [35]. Recently, new strategies have been explored to design for this space and overcome these challenges [36] [23] [32] [20]. These works show that people with dementia are capable of being involved in research when they are treated as equal partners, adequately prepared, and feel socially connected [36]. Here, it is imperative that the method is suitably adapted [31] to pose questions in an appropriate manner through which researchers can better understand and construct the experiences of dementia [8].

1.3 Technology and Design for Social Health

Social health is one of the essential components of general well-being. Needless to say, it is highly depended on communication skills, which can be compromised in the progression of dementia [37]. As a consequence, people with dementia face an increased risk of loneliness and social withdrawal. In recent years, information and communication technologies (ICT) have been used to facilitate independence, social inclusion, personal autonomy and to improve their quality of life [38].

Researchers have attempted to increase the communication competence of people with dementia, e.g. through the implementation of augmentative and alternative communication (AAC) strategies. These are designed to capitalize on the remaining strengths of individuals and take advantage of environmental cues. Bourgeois [39] explored the use of "memory wallets" assisting in the retrieval of personal information and facilitating conversation. The adoption of this simple external aid was found to prompt a two-way communication.

Communication issues may also arise from the lack of knowledge of caregivers and family members on how to interact with people with dementia. As a consequence, they often become reluctant to engage in conversations. This problem was addressed by Welsh et al. [30] through the development of an application Ticket to Talk designed to stimulate talks between young people and older adults with dementia. The application provides life context of the person with dementia by collecting photographs, videos, and sound clips related to significant events from the past. They serve as communication prompts and conversation starters. In a similar vein, CIRCA (Computer Interactive Reminiscence and Communication Aid) system aims at improving communication between people with Alzheimer's disease and their caregivers [40]. It comprises a multimedia database and allows the user to freely explore a variety of photographs, videos, music and textual content. The idea was built on the assumption that providing a system like CIRCA will minimize the conversational working memory load of people with dementia. Inspired by this work, Smith et al. [41] developed a digital memory aid, called multimedia biographies, in which a collection of digital media assets were used to tell the story of a persons' life. This carefully chosen media contents provide space for reminiscence and communication.

Other work found that using technology can enable people with dementia to engage in new forms of self-expressions and connect them with others in a shared social engagement [18] [42] [27] [17]. Even though technology is being seen as an effective vehicle for people with dementia contributing to their social health and independence, its use and applicability among this group remains scarce [43].

1.4 Dementia Dynamics in Design

Research highlights the importance of tailoring ICT interventions that promote social participation to specific individual needs and capacities [6] [44]. For technology contributing to social health, it is necessary to consider the utility of it appropriately to the progression and changing needs of individuals with dementia. Thereby considering also specific requirements of the circle of care around them, made up of their important others (i.e., partner or family members) [45]. The study reported in this paper is conducted as part of a larger ongoing research project. In this project, participants with dementia are involved and being treated as equal contributors in the design of technology for improving their social participation. To this end, a research-through-design [46] approach is adopted, adhering to an iterative design process covering exploration, design and evaluation [47] to address challenges in using social technology hardware such as mobile phones, computers, and tablets over time and enable social participation by designing new social technologies.

As the role of technology aiming to encourage and enrich social participation is still little explored [17] [44] [27], the aim of this paper is to build a base for a comprehensive understanding of social challenges experienced by people with dementia. The main goal of this study is to learn how their social participation can be hindered and facilitated by social technologies and find new design opportunities. This can eventually lead to increased meaningful engagement of people with dementia and enrich their social experience through technology.

2 Method

This paper reports on our exploratory field study in which we co-created design opportunities for social technology in the context of dementia. We employed a guided co-creation session to understand the needs, barriers and facilitators people with dementia experience in performing social interactions through social technology. We conducted one co-creation session per participant, that centered around three co-creative methods.

2.1 Social Interaction Within the Context of Dementia

Being aware of the challenges of including people with dementia in a participatory design process [48], a co-creation session was conducted at home to foster an empathetic relationship [49]. Following [50], we made a distinction between the person with dementia as an informant and as co-designer. To ensure that we approached participants from a collaborative design perspective, and thus involved them as co-creators, we have devised three generative, co-creative methods. The person with dementia is not seen as an 'informant' but rather as a contributor in the project [51]. This setting aimed to support the participants with dementia to take on a more active role and encourage them to share their daily experiences regarding social interactions and existing technology. Previously conducted qualitative interview study[1] provided us with insights that were used to personalize each method and ask specific questions about identified individual patterns, experiences and needs.

2.2 Participants and Setting

This study was part of a larger ongoing project. Participants were recruited through caregivers from home care institutions in the Netherlands. In total, four participants with dementia living at home together with their informal caregiver (e.g., spouse) participated. All participants with dementia had previous experience with using technology, either independently or with support. The participants shared the same socio-cultural context. Pseudonyms are used throughout for privacy purposes. We conducted four co-creation sessions, each involving a participant with dementia, accompanied by a spouse.

Prior to the co-creation session, the researchers sent a letter to potential participants and followed up with a phone call to ask for participation. The following inclusion criteria were adhered to: (1) participant is diagnosed with early-stage dementia caused by Alzheimer's Disease or Vascular dementia, indicated by a care professional; (2) participant lives at home together with an informal caregiver (e.g. spouse); (3) participant is willing to participate; (4) participant has no visual or auditory processing disorder and has sufficient physical ability to participate in the co-creation session (see Table 1).

The co-creation session was conducted at people's homes. Before the beginning of a session, the researcher introduced herself and

[1]Manuscript in preparation.

Table 1. Participant information; PwD—person with dementia; CG—informal caregiver

PN	Name	Age	Diagnosis
PwD1	**Charles**	**76**	**Moderate, Vascular Dementia**
CG1	*Joanna*		
PwD2	**Frank**	**79**	**Mild, Alzheimer's disease**
CG2	*Sarah*		
PwD3	**Julia**	**87**	**Moderate, Alzheimer's disease**
CG3	*Michael*		
PwD4	**Monica**	**76**	**Moderate, Alzheimer's disease**
CG4	*Sam*		

engaged in a brief conversation to put participants at ease [52]. Subsequently, the researcher explained the purpose and the course of actions of the session. The participants were informed about their rights to withdraw at any time during the session. Finally, the spouse of the participant with dementia was asked to supplement the conversation only when necessary so that the participant with dementia was able to express his or her feelings and thoughts.

During the co-creation session the principal researcher guided the activities in order to go through the co-creative methods (see Figs. 1 and 2). The generative tools were placed between the principal investigator and the couple to stimulate dialogue. By using these tools as building blocks, we aimed to encourage the participants with dementia to steer the conversation. As follows, it allows the researcher to get a better understanding of the role of the context in hindering or facilitating the use of social technology, and to extract their experiences and perspectives to steer the design agenda for social technologies.

1. Social activity and leisure time canvas. To facilitate a meaningful conversation with people with dementia, a social activity and leisure time canvas was developed, inspired by [53]. In line with this research, we created a modified canvas to gain insights into solely the social and leisure time activities that the participants still pursue, would like to pursue more often and can no longer pursue together with their significant others (see Fig. 1a). The main objective was to allow participants to reflect on barriers and facilitators in performing social activities, and share whether social technologies were used or not. Social activities and leisure time cards were used to distribute them among the different columns and get the conversation going. These cards were personalized for each participant, based on the knowledge gathered during previous research activities with the participants. Blank cards were used to include personal activities that were not previously identified.

2. Technology mapping canvas. Secondly, we developed a technology mapping canvas to explore the context of use of current social technologies (such as a phone, tablet or computer) and the social and environmental arrangements in which these are situated to fulfill their social needs (see Fig. 1b). This empathy tool consists of a canvas on which the researcher could draw a map of placement of the technology in use. In parallel to this activity, questions were asked in a sensitive and non-intrusive way to preserve a safe atmosphere.

3. Experience diary. Finally, we developed an experience diary to give voice to everyday experiences and contexts of people with dementia [54], both positive and negative, regarding social interactions and the use of social technology (see Fig. 2). After a brief introduction, it was left with the participants for two weeks. Diary studies are known to provide rich insights about personal motives and feelings, and can provide a decisive understanding into the daily lives of participants with dementia when suitably adapted [55] [31]. Inspired by this literature, we developed a modified experience

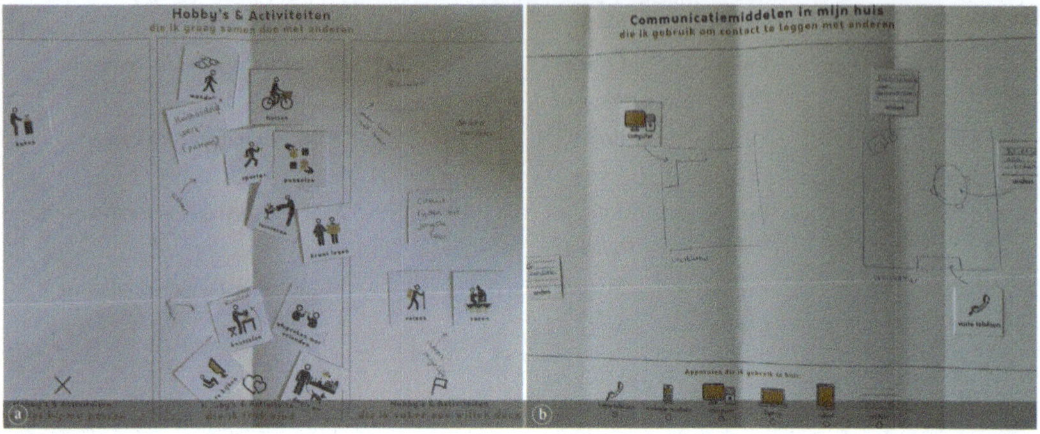

Fig. 1. Two co-creative canvases with (**a**) a social activity and leisure time canvas, and (**b**) a technology mapping canvas

Fig. 2. The third co-creative tool with an experience diary consisting of three categories of personalized question cards, a dice, a camera, a voice recorder and a notebook

diary for each couple that suited our research question. The diary contained personalized question cards that were mainly directed towards the participant with dementia. These questions were based on the insights from earlier findings and divided into three categories of cards: (1) 'my daily life' cards (e.g., "What would you like to do today?"), (2) 'my social life' cards (e.g., "What kind of activity would you like to do with others at this moment?"), and (3) 'my device usage' cards (e.g., "What device did you use today to connect with others?"). Open questions were used to stimulate a conversation between the couples. A dice was employed in the diary study to ensure randomness in card selection. Each couple was asked to roll the dice at least twice a day—which added a playful element to the research activity [31]—and thus answer at least two questions a day. They were free to choose when and where to sit together for the experience diary, and whether they preferred to answer the questions using a notebook, voice recorder or a camera [55].

2.3 Ethics

This research was approved by the Ethics Review Board of Tilburg School of Social and Behavioral Sciences (number EC-2018.76).

Written consent was given by all participants themselves after the principal researcher went through the information letter with them. The principal researcher conducting the in-context sessions had previous experience with similar sessions with people with dementia. During the session, the participants were repeatedly asked if they would like to continue or preferred to stop. The researcher paid attention to the bodily expressions of the person with dementia that would be indicative of the need to interrupt or terminate the session [11]. A reciprocal approach was employed to lower the burden on the participants and to enable them to feel comfortable [56] in an open and safe setup with careful attention to their needs [57].

2.4 Data Collection and Analysis

Data collected in this study consisted of (1) audio recordings of the sessions, (2) photographic documentation of the completed co-creative tools, and (3) the documentation of four diaries. All audio recordings were deleted once the transcripts were made. All data were analyzed by a thematic analysis in ATLAS/ti software, which involved coding, interpretation, and generation of themes and sub-themes [58]. The transcripts, documentation and evolving theme list were discussed by four experienced qualitative researchers. We conducted a bottom-up analysis to identify participants' needs, barriers they encounter in engaging into social interactions, difficulties in using existing social technology, and strategies that are being used by the participants or their families to mitigate the influence of dementia on social interactions. The quotations used within this article are translated from Dutch to English.

3 Results

Our thematic analysis of data gathered from the co-creation sessions centers around four main themes allowing us to identify design opportunities for improving social interactions of people with dementia through social technology. The main selected themes are: (1) Social needs of people with dementia living self-reliant, (2) Barriers in engaging in social interactions/activities, (3) Strategies used by people with dementia and their families to mitigate the impact of the condition on social interactions, (4) Difficulties in using existing social technological solutions. We carefully examine the themes and sub-themes in the following sections.

3.1 Social Needs of People with Dementia that Live Self-reliant

In order to explore the design opportunities for new social technologies that would draw on the experience of people with dementia and—subsequently—enrich that experience, we started our inquiry with discovering the most commonly expressed needs of self-reliant people with dementia involved in the study. Through the responses given by the participants and their relatives, three main themes kept recurring: the need for *fulfilling relationships, agency* and *shared activities*.

Fulfilling relationships. Reviewing social contacts of people with dementia, allowed us to identify three important components influencing the perception of individual relationships as satisfactory and fulfilling.

The need for ***close interactions*** comes before anything else. Staying in touch with family members and close friends has been indicated as an essential component of satisfactory social life by the participants. When asked about the most valuable aspect of their current social context and how it has changed recently with respect to the past, one of the couples wrote down in the diary "*Contact with children, grandchildren, direct neighbors and good acquaintances. The quantity decreases slightly, while the quality improves.*" [*Julia, Michael*]. Similarly, when asked by the researcher about the concept of an ideal social life another participating couple said:

Researcher: What is an example of an ideal social life for you right now? Joanna: Spending time together with our children and grandchildren. That is important to us. They are regarded as the number one. Charles: Yes, logically.

In addition, a broader context of ***social presence*** has been indicated as a relevant part of social well-being. For people with dementia participation in social activities, being involved in the life of local communities and having various social ties appears to be an important component of social health. One of the participants particularly enjoyed home visits of her husband's business clients, as she has seen it as an opportunity to have a little chat with the visitors. Another person with dementia expressed how participating in the meetings of local community (neighbourhood association) helps her to remain connected to other people, emphasizing that these social ties are the main reason why she is still a part of the group.

Apart from the nature of relationships and the proximity of social networks, the way in which a person with dementia stays connected to the social environment should be taken into consideration. According to our data, the most desirable way of connecting to others was through having ***mundane conversations***, simple everyday chit-chats. When talking about interacting with other people in a broader social context, one of the participants said cheerfully: *"I have more contact with people. I don't really have much trouble with it. It's just, you go somewhere, no matter if it's morning, afternoon or evening: good morning everyone, how are you here? Well, and then people start talking." [Monica].*

Agency. The importance of the sense of agency in dementia has been confirmed in our design inquiry. While often the assistance of family members or carers eventually becomes indispensable, being devoid of simple everyday choices may lead the person with dementia to the feeling of loss of control and frustration. The need for agency manifests different per person. Some of the participants were clearly expressing the need of playing an instrumental role in carrying out activities. Monica said: *"I can still pay my bills on the computer. I know how to do it, and I do it as it should be. But my husband then says: wait until the children are with us. Then they can watch if I am doing it right. My husband does not know how to use the computer. I keep saying to myself: I can still do it. But he doesn't fully trust me."* In other cases, the focus was not on the capacity to act, but rather on taking an active role in the decision-making process. It is particularly prominent when talking about stepping back from previous chores and responsibilities or social circles. In another example from Monica, being a part of a neighborhood association, the possibility to make a choice of when to resign from her membership appeared to be of great importance.

Shared activities. Lastly in this section, participants expressed the desire to stay engaged in shared activities, carry on with simple everyday tasks, chores and personal interests. When asked by a researcher what are the things that she would like to continue doing, Monica answered:

Monica: ... what I would like to do with someone else, is to play card games and just having chit chats. As we do right now, drinking a cup of coffee and having pleasant chats. My husband is not a big fan of playing cards, but I would really like to join a group to play cards one evening.

Another important component of well-being, falling within the category of a need of sharing activities with others, appears to be physical activity. Taking part in a variety of group sports activities, going for a walk or cycling has been identified as a relevant and enjoyable part of a daily routine of our participants.

3.2 Barriers in Engaging in Social Interactions or Activities

We identified two main categories of barriers in engaging in social interactions: *external* and *internal* conditions which will be explained in the following sections.

External barriers. Three main aspects of the external environment have been identified as impeding social inclusion of people with dementia: ***environmental conditions***, ***location/distance***, and ***presence and behavior of***

others. The impact of environmental factors (such as weather) or the distance between person's with dementia household and a given meeting point on their participation in social activities is fairly straightforward. In addition, an important external factor influencing people's social health is not only their physical, but also social environment. The absence of family or friends, or their lack of availability and willingness to engage in interactions, may preclude the person with dementia from building meaningful connections.

Internal barriers. The most prominent message deriving from the data collected was decreasing self-confidence of the participants with dementia. We have identified two subthemes being the most relevant internal factors leading to social withdrawal or insecurity in the social context: ***fear of decreasing social competence*** and ***perceived loss of technical competence***.

Fear of decreasing social competence. As a result of deteriorating cognitive capacities, especially memory and attention difficulties, people with dementia experience the sense of inadequacy in the social context. Over the course of the co-creation sessions, the problem of fearing to engage or initiate social contact kept reappearing. The participants confessed being worried about their forgetfulness, loosing track of the conversation and possible awkward behavior. It appears to be particularly difficult for people with dementia to interact with others through digital means, i.e. phone calls. Charles expressed explicitly his reluctance to communicate in this way:

Charles: She sometimes says to me, "You have to call." But I hate it to make outgoing calls, because during the conversation I cannot get the answer and then I do not know it anymore.

Perceived loss of technical competence. Despite procedural skills being fairly well preserved in people with dementia [20] [25], because of experienced deterioration of cognitive abilities, their subjective perception of technical and instrumental competency might be compromised. As a result of supposed inability, they tend to give up on the attempts to interact with new technologies and step back from various commitments and activities they used to enjoy. Oftentimes, the reluctance to use technology impedes maintaining regular contacts with others who, for example, do not live in a close proximity of the person with dementia. Abandonment of interests or giving up on previous responsibilities, on the other hand, lessens the chances of meeting new people and maintaining previous circles of friends and acquaintances. All of that together may lead to deepening the sense of social isolation of people with dementia.

3.3 Strategies Used by People with Dementia and Their Families to Mitigate the Impact of the Condition on Social Interactions

Participants have already implemented some simple strategies to mitigate the negative impact of the condition on their everyday lives and social interactions. Identifying these means can provide valuable insights into how future social technologies should be designed. Based on the collected data, we distinguished three main strategies facilitating living with dementia and maintaining social connectedness: *appropriate levels of assistance, routine,* and *self-management strategies.*

Appropriate levels of assistance. With cognitive and functional decline following the progression of dementia, people affected by it will need assistance over time. However, in order to prompt activity among people with dementia and promote their independence for as long as possible, it is essential to offer an appropriate level of assistance, instead of taking over tasks from them in all life domains. Based on the insights gathered from our participants, we identified three points on the support spectrum: ***prompts***, ***supported autonomy***, and ***inclusive takeover***.

The first, lowest level of assistance, aims to maintain full autonomy and independence of the person with dementia. The support takes the form of *prompts* coming from a system or device. The prompts may trigger either the execution of a particular action (e.g. answering the incoming call) or recollection of memories (e.g. exposing the person with dementia to a video that leads to recollection of old memories and prompts a conversation) leading to a start of a conversation.

The second level of assistance—*supported autonomy*—allows the person with dementia to maintain control over a given task or interaction, yet with verbal or non-verbal cues from the side of a partner or family members and under their supervision. It might be while the person with dementia tries to send a message or an email with a guidance of a significant other who, in case of necessity, suggests the next steps required to complete the action or indicates which buttons to press.

At the end of the support spectrum there is the action of *inclusive taking over* which entails the person with dementia to be indirectly involved in the execution of a given action, that is done by a family member. As an example of this dynamics, illustrated by one of the couples, it was the informal caregiver (partner) of the person with dementia who received all of the messages and pictures from the family members. Afterwards, he presented the content to his wife and sent back a response on behalf of both of them. It enabled the person with dementia to have a feeling of staying involved, without the burden of using technological solutions that might have become too demanding for her to use.

Routine. Another strategy facilitating social interactions is incorporating elements of routines. Having regular group meetings, coffee time with friends, or gym appointments helps to make social interactions an integral part of the day. For example, Monica combines physical activities with socializing by having a fixed appointment in a swimming pool with her friend. She explains: *"This is usually agreed, but normally if we do not call each other, the appointment will remain on Wednesday evening."*. Moreover, incorporating routines into everyday lives of people with dementia may help to circumvent the experienced memory challenges.

Self-management strategies. The third category in this section are self-management strategies. We identified three approaches within this category: ***external prompts***, ***mnemonics***, and ***masking***.

A common tactic used by people with dementia and their close ones is using *external memory prompts*. Some of the involved participants used them as a communication means, for example family members leave short messages stating where they went. In another example, a participant with dementia kept a journal helping himself to find answers to frequently repeated questions or remember events of the passing days:

> *Sarah: If he asks me three times the same thing, he writes it down in his journal. He writes down all important things.*
> *Frank: Yes, I can always read it back.*

Mnemonic strategies are used as a technique aiming at memory improvement. They are based on creating connections between prior knowledge and new information through the use of various types of cues, i.e. images, key words, rhyming words, acronyms. Using *mnemonic strategies* is more cognitively demanding than external memory prompts and turned out to be less common among our participants with dementia. However, it appears to be helpful for some of the participants to make use of word similarities, humorous associations or acronyms in order to recall certain information, i.e. people's names.

The last strategy commonly implemented by people with dementia themselves is *masking* their cognitive difficulties. Revealing the diagnosis to one's social circle might be difficult, as the term *dementia* still carries a social stigma. It arose from the co-creation sessions that some of the participants try to conceal from their surroundings that they experience cognitive decline.

3.4 Difficulties in Using Existing Social Technologies

We learned what the most prominent barriers and difficulties where while using social technologies. The issues that have been mentioned most frequently in this part of our inquiry were: *placement* and *level of complexity*.

Placement. The placement of technological devices appears to have a very important and multidimensional role. On one hand, it can be seen as a means of facilitating habitual usage of a system. If a device is placed in a fixed, easily available location, it might serve as a prompt for the person with dementia to use it and it potentially eliminates the problem of not being able to find it. On the other hand, if a device is being stored in a hidden place, locating it becomes a challenge leading to diminished use. It has been a commonly reported problem with, for example, mobile phones. They struggled with both finding it and charging it. This example points out to a broader problem of forgetfulness that can be amplified by a poor choice of the location of a device. If it is not placed in sight charging or bringing technology along when they leave the house might become challenging.

Complexity in existing technologies. These days there is not only a wide range of technological solutions on the market, but they also tend to keep being replaced and modified at staggering rates. It might be difficult to keep up with new technologies even for younger generations with high computer literacy, meaning this is even more demanding for older adults with lower digital literacy. In our co-creation sessions, we found evidence for three main features of new technologies that make them being perceived as too complex by people with dementia: ***confusing design***, ***fast developments***, and ***time pressure***. With rapidly changing interfaces, a wide range of new functionalities and User Interfaces that are not always clear and intuitive for older adults, completing tasks that require taking prompt actions, i.e. online banking, result in being stressful and difficult. Monica says: *"Technology is moving rapidly. I need my husband and children for help or support."*

3.5 Initiative

Ultimately, we found that the role of initiative is particularly important and complex in maintaining social interactions by people with dementia. This factor has not been described within any of the above-mentioned themes, as it appears to be in a special, two-way relationship with some of the other variables. On one hand, due to previously described internal barriers experienced by people with dementia, they less frequently take the lead or initiative to engage in social contact. They do not feel confident enough, they are afraid of their behavior being perceived as inadequate or awkward, thus they tend to avoid initiating interactions. On the other hand, the more they withdraw from their social circles, the fewer opportunities they have to challenge themselves, take the lead and recognize the abilities that remained intact. Consequently, prompting the initiative of people with dementia can be seen as an element of strategies mitigating the impact of the condition on social interactions.

4 Discussion

Our findings provide insights into the social abilities and limitations people with dementia experience, and the nature of interaction with existing social technology. By using co-creative tools, opportunities for design to enrich their social experiences have been co-constructed with people with dementia and their spouses. Ultimately, we provide a list of practical implications for design.

4.1 Design Considerations for Social Technology

The above findings contribute to existing research in social technology and dementia by providing an extensive insight into the social experiences and interactions of people with dementia in their daily home environment. In the following section, we will further outline the considerations for the design in this context.

4.1.1 Compensating for Internal and External Barriers Experienced by People with Dementia Through Design

The co-creation outcomes indicated that internal barriers as experienced by people with dementia cause a decrease in self-confidence in interacting with significant others. As a consequence, this may lead to social withdrawal. To enhance person's with dementia self-confidence, it is worthwhile looking at what role design can play in circumventing the internal barriers [59] and stimulating the desired social participation. One possible solution would be to design technologies that give people with dementia the feeling of being in control and reducing the distance [60]. Technology design can provide support in anchoring conversations. One possible approach is to let people with dementia interact with physical prompts as a way to indicate intentions for social sharing [17]. The involvement of the social circle (i.e., family members or significant others) is essential to trigger these intentions and to support agentic abilities of the person [27] to get the conversation started [30]. On the other hand, building on the concept of synchronous social events making it dependent on others, it is also profitable looking at a more hybrid approach, in which the technology itself could trigger social intentions asynchronously.

According to our participants, external barriers can hamper social participation. Existing social technology can help bridge, for example, the geographical distance between people. Yet, at the same time, it can create new internal barriers for individuals who lack technical expertise or necessary hardware [61]. We can contribute to the latter by creating accessible designs for people with dementia, matching the needs and capacities of the user with dementia [28].

4.1.2 Enhancing Agency Through Social and Physical Environment

Within HCI, a strong relationship between technology and agency was reported in the literature [27] [18]. Agency is created and maintained through the dynamics between individuals, those with dementia, and those without. As demonstrated in our findings, the manifestation of agency is different for each individual with dementia. This is closely related to the facilitating strategies we found used by people with dementia to maintain their agentic abilities when interacting with existing social technology. For example, for all of our participants, it seems natural to support each other and to be there for each other within a social circle. As a result, the agency of a person with dementia is not only co-shaped but can also be strengthened by the environment. In order to improve social participation among people with dementia it is, therefore, recommended looking at the construction of everyone's agency, taking into account the social circle around it, and to enhance it through design [30] [42] [17].

In addition to these social dynamics, the physical environment can also provide a potential strategy. This aligns with work by [27], in which physical cues remind people with dementia to perform an action. Our findings extend current understandings and show that people with dementia adapt their environment with tangible prompts supporting them in interaction with technology. Building on these used adaptation strategies can help designers to comprehensively support people with dementia through design, specifically when it comes to interaction with social technology.

4.1.3 Learning from Previous Experiences with Existing Social Technologies

Our findings pinpoint to the aspects that are positive and negative about existing social technologies. The properties of technology contribute to whether it is used or not. Bringing the current mechanisms of use to the center of design can better support people with dementia in performing social interactions through technology. Drawing on their preserved abilities, technology design can help to experience conversations as pleasant again. However, a certain level of flexibility and open appropriation in the design is necessary to make the technology suitable for different scenarios and contexts [22], and to foreground the social experiences of people with dementia.

4.2 Recommendations for Future Research in Social Technology for People with Dementia

Based on the obtained results, we present possible future directions and implications for research in the area of social technology in the context of dementia.

4.2.1 Embracing the Importance of a Broader Social Context

Our findings point to the importance of social interactions going beyond the people closest to individuals with dementia. Even though, as one would expect, strong social ties and close personal relationships are considered highly valuable, our participants also revealed to appreciate the presence of others, everyday chitchats with strangers and the sense of connectedness to the community. Building dementia-friendly, safe and open communities can be of a great challenge [62], yet it can be a challenge addressed by future social technology designs. Design researchers should reflect on how to engage local communities in supporting people with dementia to stay involved and active, without compromising their safety and see the role of design beyond the technology. Initiating a dialogue with the local services and communities seeking to explore and understand their perspective is vital, so as to create more inclusive social space for people with dementia.

4.2.2 Sharing Meaningful Activities

Having a fulfilling social life is composed of a variety of factors and may take different forms, depending on individual preferences and circumstances. It is easy to fall into a trap of thinking that with communication abilities being affected incrementally by the progression of dementia, meaningful interactions become out of reach. However, it is to bear in mind that communication can occur in many different shapes and forms and that meaningful interactions go beyond verbal expressions [63]. People with dementia involved in the study expressed their desire to carry on with their hobbies, physical activities and daily chores. Incorporating shared activities, promoting the concept of doing things together, and providing a safe space for expressing the wish of finding a companion for a particular activity could be a way of facilitating social engagement for people with dementia through design.

4.2.3 Incorporating Social Contact into the Everyday Routines

As stated in the results section, taking the initiative is an ability that decreases with the progression of dementia. Setting up a meeting or making a phone call of one's own accord happens with a diminishing frequency, yet often, while exposed to a social situation, people with dementia decide to engage in the interaction and prove to appreciate it. We have seen that, in many cases, making social experience a part of daily routines can partially compensate for the decreased initiative and prevent the exclusion of people with dementia as their condition progresses. Technology can play a facilitating role in establishing these routines. To this end, we found that external prompts can help maintain, facilitate and create social interactions. By bringing strategies—already used by people with dementia and their families—to the center of technology design, we can promote building new routines centered around social health and connections. Future designs should take into account the necessity of embodying mitigation strategies reducing the impact of external and internal barriers on staying socially involved over the course of dementia.

5 Conclusion

In this paper, we present the insights arising from co-creation sessions conducted with four people with dementia and their spouses focusing on social health and communication technologies. Based on the outcomes of these participatory research activities, we were able to define the main limitations and difficulties experienced by our participants in the context of social interactions and technology usage, and explore what abilities and compensating strategies may be helpful in circumventing these barriers.

Through engaging people with dementia and their partners in a participatory way, we have gathered insights into their individual experiences of the social and technological world. Building further on their needs and capacities, we identified design opportunities for improving social health through communication technology. In our design considerations, we highlight the importance of: (1) compensating for internal and external barriers experienced by people with dementia through design, (2) enhancing agency through the social and physical environment, (3) learning from previous experiences with existing social technologies.

Ultimately, we present possible design implications and future research directions. We point to the importance of reflecting not only on the role of people affected by dementia in building and maintaining their personal relationships but also on the impact of their social circles and local communities. We call for design to prompt the engagement of social networks of people with dementia and promote the sense of connectedness through shared meaningful activities. Last but not least, we encourage designers to learn from the strategies spontaneously implemented by people with dementia and their environment to compensate for the difficulties caused by the progressing cognitive decline.

Acknowledgements We would like to thank all participants that shared their experiences during the study. We are grateful to ZonMW for funding our 'Dementia Dynamics in Design' project in the Create Health program, a multi-disciplinary consortium consisting of researchers from the Department of Industrial Design of the Eindhoven University of Technology, the Tilburg School of Social and Behavioral Sciences, a Mental Care institute (GGzE), and an innovation network for active and healthy ageing (Corporation Slimmer Leven 2020).

References

1. Livingston, G., Sommerlad, A., Orgeta, V., Costafreda, S.G., Huntley, J., Ames, D., Ballard, C., Banerjee, S., Burns, A., Cohen-Mansfield, J., Cooper, C., Fox, N., Gitlin, L.N., Howard, R., Kales, H.C., Larson, E.B., Ritchie, K., Rockwood, K., Sampson, E.L., Samus, Q., Schneider, L.S., Selbæk, G., Teri, L., Mukadam, N.: Dementia prevention, intervention, and care (Dec 2017). https://doi.org/10.1016/S0140-6736(17)31363-6
2. Gaggi, O., Kolasinska, A., Palazzi, C.E., Quadrio, G.: Safety first? users' perception of wearable sensor networks for aging. Mob. Netw. Appl., 1–9 (Feb 2019). https://doi.org/10.1007/s11036-019-01234-6
3. Derksen, E., Vernooij-Dassen, M., Gillissen, F., Olde Rikkert, M., Scheltens, P.: Impact of diagnostic disclosure in dementia on patients and carers: qualitative case series analysis. Aging & Ment. Health **10**(5), 525–531 (Sep 2006). https://doi.org/10.1080/13607860600638024
4. Tranvåg, O., Petersen, K.A., Nåden, D.: Relational interactions preserving dignity experience: Perceptions of persons living with dementia. Nurs. Ethics 22(5), 577–593 (Aug 2015). https://doi.org/10.1177/0969733014549882
5. Van Der Roest, H.G., Meiland, F.J., Comijs, H.C., Derksen, E., Jansen, A.P., Van Hout, H.P., Jonker, C., Dröes, R.M.: What do community-dwelling people with dementia need? a survey of those who are known to care and welfare services. Int. Psychogeriatr. **21**(5), 949–965 (2009). https://doi.org/10.1017/S1041610209990147
6. Dröes, R.M., Chattat, R., Diaz, A., Gove, D., Graff, M., Murphy, K., Verbeek, H., Vernooij-Dassen, M., Clare, L., Johannessen, A., Roes, M., Verhey, F., Charras, K., van Audenhove, C., Casey, D., Evans, S., Fabbo, A., Franco, M., Gerritsen, D., Vittoria Gianelli, M., Gonalves-Pereira, M., Gzil, F., van Hout, H., Innes, A., Hee Jeon, Y., Koopmans, R., Kristensen, F., Losada Baltar, A., McEvoy, P., McHugh, J., Meiland, F., Moniz-Cook, E., Parkes, J., Rymaszewska, J., Spruytte, N., Surr, C., de Vugt, M., Wolf-Ostermann, K., Zuidema, S.: Social health and dementia: a European consensus on the operationalization of the concept and directions for research and practice. Aging Ment. Health **21**(1), 4–17 (2017). https://doi.org/10.1080/13607863.2016.1254596
7. Webb, J.: Conversation takes two: understanding interactions with people with dementia. Disabil. Soc. 32(7), 1102–1106 (Aug 2017). https://doi.org/10.1080/09687599.2017.1321225
8. Wallace, J., Wright, P.C., McCarthy, J., Green, D.P., Thomas, J., Olivier, P.: A design-led inquiry into personhood in dementia. In: Proceedings of the SIGCHI Conference on Human Factors in Computing Systems—CHI '13, p. 2617. ACM Press, New York, New York, USA (2013). https://doi.org/10.1145/2470654.2481363
9. Kitwood, T., Bredin, K.: Towards a theory of dementia care: personhood and well-being. Ageing Soc. **12**(3), 269–287 (1992). https://doi.org/10.1017/S0144686X0000502X
10. Sabat, S.R., Harré, R.: The construction and deconstruction of self in alzheimer's disease. Ageing Soc. **12**(4), 443–461 (1992). https://doi.org/10.1017/S0144686X00005262

11. Dewing, J.: Process consent and research with older persons living with dementia. Res. Ethics **4**(2), 59–64 (2008). https://doi.org/10.1177/174701610800400205
12. Kitwood, T.: Dementia reconsidered: the person comes first. Open University Press, Buckingham (1997)
13. McCormack, B., Reed, J.: Editorial: 'evidence-based healthcare—a lot of bull?'. Int. J. Older People Nurs. **1**(3), 129–130 (sep 2006). https://doi.org/10.1111/j.1748-3743.2006.00035.x
14. Constant, E., Pickering, A.: The mangle of practice: time, agency, and science. Technol. Cult. **38**(3), 815 (Jul 1997). https://doi.org/10.2307/3106908
15. Suchman, L.: Agencies in technology design: feminist reconfigurations. In: Online Proceedings of the 5th European Symposium on Gender ICT
16. Tholander, J., Normark, M., Rossitto, C.: Understanding agency in interaction design materials. In: Conference on Human Factors in Computing Systems—Proceedings, pp. 2499–2508. ACM Press, New York, New York, USA (2012). https://doi.org/10.1145/2207676.2208417
17. Lazar, A., Edasis, C., Piper, A.M.: Supporting people with dementia in digital social sharing. In: Conference on Human Factors in Computing Systems—Proceedings, vol. 2017–May, pp. 2149–2162. Association for Computing Machinery, New York, NY, USA (may 2017). https://doi.org/10.1145/3025453.3025586
18. Lazar, A., Edasis, C., Piper, A.M.: A critical lens on dementia and design in HCI. In: Proceedings of the 2017 CHI Conference on Human Factors in Computing Systems—CHI '17. pp. 2175–2188. ACM Press (2017). https://doi.org/10.1145/3025453.3025522
19. Hodge, J., Balaam, M., Hastings, S., Morrissey, K.: Exploring the design of tailored virtual reality experiences for people with dementia. In: Conference on Human Factors in Computing Systems—Proceedings, vol. 2018–April, pp. 1–13. Association for Computing Machinery, New York, New York, USA (Apr 2018). https://doi.org/10.1145/3173574.3174088
20. Houben, M., Brankaert, R., Bakker, S., Kenning, G., Bongers, I., Eggen, B.: Foregrounding everyday sounds in dementia. In: DIS 2019—Proceedings of the 2019 ACM Designing Interactive Systems Conference, pp. 71–83. Association for Computing Machinery, Inc, New York, New York, USA (Jun 2019). https://doi.org/10.1145/3322276.3322287
21. Morrissey, K., McCarthy, J., Pantidi, N.: The value of experience-centred design approaches in dementia research contexts. In: Proceedings of the 2017 CHI Conference on Human Factors in Computing Systems—CHI '17, pp. 1326–1338. ACM Press, New York, New York, USA (2017). https://doi.org/10.1145/3025453.3025527
22. Thoolen, M., Brankaert, R., Lu, Y.: AmbientEcho: exploring interactive media experiences in the context of residential dementia care. In: Proceedings of the 2020 ACM Designing Interactive Systems Conference. pp. 1495–1508. ACM, New York, NY, USA (Jul 2020). https://doi.org/10.1145/3357236.3395432
23. Branco, R.M., Quental, J., Ribeiro, Ó.: Personalised participation: an approach to involve people with dementia and their families in a participatory design project. CoDesign **13**(2), 127–143 (2017). https://doi.org/10.1080/15710882.2017.1310903
24. Hendriks, N., Huybrechts, L., Slegers, K., Wilkinson, A.: Valuing implicit decision-making in participatory design: a relational approach in design with people with dementia. Des. Stud. **59**, 58–76 (2018). https://doi.org/10.1016/j.destud.2018.06.001
25. Morrissey, K., Wood, G., Green, D., Pantidi, N., McCarthy, J.: 'I'm a rambler, I'm a gambler, I'm a long way from home': the place of props, music, and design in dementia care. In: Proceedings of the 2016 ACM Conference on Designing Interactive Systems—DIS '16, vol. 32, pp. 1008–1020. ACM Press, New York, New York, USA (2016). https://doi.org/10.1145/2901790.2901798
26. Lazar, A., Edasis, C., Piper, A.M.: A critical lens on dementia and design in HCI. In: Proceedings of the 2017 CHI Conference on Human Factors in Computing Systems—CHI '17, pp. 2175–2188. ACM Press (2017). https://doi.org/10.1145/3025453.3025522
27. Foley, S., Welsh, D., Pantidi, N., Morrissey, K., Nappey, T., McCarthy, J.: Printer pals: experience-centered design to support agency for people with dementia. Proceedings of the 2019 CHI Conference on Human Factors in Computing Systems—CHI '19, pp. 1–13 (2019). https://doi.org/10.1145/3290605.3300634
28. Thoolen, M., Brankaert, R., Lu, Y.: Sentic: A tailored interface design for people with dementia to access music. In: Companion publication of the 2019 on designing interactive systems conference 2019 companion—DIS '19 Companion, pp. 57–60. ACM Press, New York, New York, USA (2019). https://doi.org/10.1145/3301019.3325152
29. Wright, P., McCarthy, J.: Experience-centered design: designers, users, and communities in Dialogue. Synth. Lect. Hum.-Cent.Ed Inform. **3**(1), 1–123 (Jan 2010). https://doi.org/10.2200/S00229ED1V01Y201003HCI009
30. Welsh, D., Morrissey, K., Sarah, F., McNaney, R., Salis, C., McCarthy, J., Vines, J.: Ticket to talk: supporting conversation between young people and people with dementia through digital media. In: Proceedings of the 2018 CHI Conference on Human Factors in Computing Systems—CHI '18 (2018). https://doi.org/10.1145/3173574.3173949
31. Suijkerbuijk, S., Brankaert, R., de Kort, Y.A., Snaphaan, L.J., den Ouden, E.: Seeing the first-person perspective in dementia: a qualitative personal evaluation game to evaluate assistive technology for people affected by dementia in the home context. Interact. Comput. **27**(1), 47–59 (jan 2015). https://doi.org/10.1093/iwc/iwu038

32. Treadaway, C., Kenning, G.: Sensor e-textiles: person centered co-design for people with late stage dementia. Work. Older People **20**(2), 76–85 (Jun 2016). https://doi.org/10.1108/WWOP-09-2015-0022
33. Meiland, F., Innes, A., Mountain, G., Robinson, L., van der Roest, H., García- Casal, J.A., Gove, D., Thyrian, J.R., Evans, S., Dröes, R.M., Kelly, F., Kurz, A., Casey, D., Szczesniak, D., Dening, T., Craven, M.P., Span, M., Felzmann, H., Tsolaki, M., Franco-Martin, M.: Technologies to support community—dwelling persons with dementia: a position paper on issues regarding development, usability, effectiveness and cost-effectiveness, deployment, and ethics. JMIR Rehabil. Assist. Technol. **4**(1), e1 (jan 2017). https://doi.org/10.2196/rehab.6376
34. Mäki, O., Topo, P.: User needs and user requirements of people with dementia: multimedia application for entertainment. Assist. Technol. Res. Ser. **24**, 61–75 (2009)
35. Astell, A.J., Ellis, M.P., Bernardi, L., Alm, N., Dye, R., Gowans, G., Campbell, J.: Using a touch screen computer to support relationships between people with dementia and caregivers. Interact. Comput. **22** (4), 267–275 (2010). https://doi.org/10.1016/j.intcom.2010.03.003
36. Bartlett, R.: Modifying the diary interview method to research the lives of people with dementia. Qual. Health Res. **22**(12), 1717–1726 (2012). https://doi.org/10.1177/1049732312462240
37. Bourgeois, M.S., Dijkstra, K., Burgio, L., Allen-Burge, R.: Memory aids as an augmentative and alternative communication strategy for nursing home residents with dementia. AAC: Augment. Altern. Commun. **17**(3), 196–210 (2001). https://doi.org/10.1080/aac.17.3.196.210
38. Pinto-Bruno, A′. C., García-Casal, J.A., Csipke, E., Jenaro-Río, C., Franco-Martín, M.: ICT-based applications to improve social health and social participation in older adults with dementia: a systematic literature review. Aging Ment. Health **21**(1), 58–65 (2017). https://doi.org/10.1080/13607863.2016.1262818
39. Bourgeois, M.S.: Enhancing conversation skills in patients with alzheimer's disease using a prosthetic memory aid. J. Appl. Behav. Anal. **23**(1), 29–42 (Mar 1990). https://doi.org/10.1901/jaba.1990.23-29
40. Astell, A., Alm, N., Gowans, G., Ellis, M., Dye, R., Campbell, J.: CIRCA: a communication prosthesis for dementia. Assist. Technol. Res. Ser. **21**, 67–76 (2008)
41. Smith, K.L., Crete-Nishihata, M., Damianakis, T., Baecker, R.M., Marziali, E.: Multimedia biographies: a reminiscence and social stimulus tool for persons with cognitive impairment. J. Technol. Hum. Serv. **27**(4), 287–306 (Nov 2009). https://doi.org/10.1080/15228830903329831
42. Hodge, J., Montague, K., Hastings, S., Morrissey, K.: Exploring media capture of meaningful experiences to support families living with dementia, pp. 1–14 (2019). https://doi.org/10.1145/3290605.3300653
43. van Boekel, Wouters, Grimberg, van der Meer, Luijkx: Perspectives of stakeholders on technology use in the care of community-living older adults with dementia: a systematic literature review. Healthcare **7**(2), 73 (2019). https://doi.org/10.3390/healthcare7020073
44. Smith, S.K., Mountain, G.A.: New forms of Information and Communication Technology (ICT) and the potential to facilitate social and leisure activity for people living with dementia. Int. J. Comput. Healthc. **1**(4), 332 (2012). https://doi.org/10.1504/IJCIH.2012.051810
45. Lorenz, K., Freddolino, P.P., Comas-Herrera, A., Knapp, M., Damant, J.: Technology-based tools and services for people with dementia and carers: mapping technology onto the dementia care pathway. Dementia **18**(2), 725–741 (2019). https://doi.org/10.1177/1471301217691617
46. Zimmerman, J., Forlizzi, J.: Research through design in HCI. In: Olson, J.S., Kellogg, W.A. (eds.) Ways of knowing in HCI, pp. 167–189 (2014). https://doi.org/10.1007/978-1-4939-0378-8
47. Brankaert, R.: Design for dementia: a design-driven living lab approach to involve people with dementia and their context. Ph.D. thesis, TU Eindhoven (2016)
48. Hendriks, N., Huybrechts, L., Wilkinson, A., Slegers, K.: Challenges in doing participatory design with people with dementia. In: Proceedings of the 13th Participatory Design Conference on Short Papers, Industry Cases, Workshop Descriptions, Doctoral Consortium papers, and Keynote abstracts—PDC '14, vol. 2, pp. 33–36. ACM Press, New York, New York, USA (2014). https://doi.org/10.1145/2662155.2662196
49. Lindsay, S., Brittain, K., Jackson, D., Ladha, C., Ladha, K., Olivier, P.: Empathy, participatory design and people with dementia. In: Proceedings of the 2012 ACM annual conference on Human Factors in Computing Systems—CHI '12. p. 521. No. March 2014, ACM Press, New York, New York, USA (2012). https://doi.org/10.1145/2207676.2207749
50. Suijkerbuijk, S., Nap, H.H., Cornelisse, L., Ijsselsteijn, W.A., De Kort, Y.A., Minkman, M.M., Baglio, F.: Active involvement of people with dementia: a systematic review of studies developing supportive technologies. J. Alzheimer's Dis. **69**(4), 1041–1065 (2019). https://doi.org/10.3233/JAD-190050
51. Sanders, E.B., Stappers, P.J.: Probes, toolkits and prototypes: three approaches to making in codesigning (2014). https://doi.org/10.1080/15710882.2014.888183
52. Nygård, L.: How can we get access to the experiences of people with dementia? Scand. J. Occup. Ther. **13**(2), 101–112 (Jan 2006). https://doi.org/10.1080/11038120600723190
53. Den Haan, M.C., Brankaert, R.G.A., Lu, Y.: The leisure time canvas. Conf. Proc. Acad. Des. Innov.

Manag. **2**(1), 4–14 (Nov 2019). https://doi.org/10.33114/adim.2019.01.421
54. Herron, R., Dansereau, L., Wrathall, M., Funk, L., Spencer, D.: Using a flexible diary method rigorously and sensitively with family carers. Qual. Health Res. **29**(7), 1004–1015 (2019). https://doi.org/10.1177/1049732318816081
55. Bartlett, R.: Modifying the diary interview method to research the lives of people with dementia. Qual. Health Res. **22**(12), 1717–1726 (2012). https://doi.org/10.1177/1049732312462240
56. Kenning, G.: Reciprocal design: inclusive design approaches for people with late stage dementia. Des. Health **2**(1), 142–162 (2018). https://doi.org/10.1080/24735132.2018.1453638
57. Hodge, J., Foley, S., Brankaert, R., Kenning, G., Lazar, A., Boger, J., Morrissey, K.: Relational, flexible, everyday: learning from ethics in dementia research. In: Proceedings of the 2020 CHI Conference on Human Factors in Computing Systems, pp. 1–16. ACM, New York, NY, USA (Apr 2020). https://doi.org/10.1145/3313831.3376627
58. Braun, V., Clarke, V.: Using thematic analysis in psychology. Qual. Res. Psychol. **3**(2), 77–101 (Jan 2006). https://doi.org/10.1191/1478088706qp063oa
59. Sanders, D., Scott, P.: Literature review: technological interventions and their impact on quality of life for people living with dementia. BMJ Health Care Inform. **27**(1), 1–10 (2020). https://doi.org/10.1136/bmjhci-2019-100064
60. Lazar, A., Thompson, H.J., Demiris, G.: Design recommendations for recreational systems involving older adults living with dementia. J. Appl. Gerontol. **37**(5), 595–619 (May 2018). https://doi.org/10.1177/0733464816643880
61. Yasuda, K., Kuwahara, N., Morimoto, K.: Remote reminiscence talking and scheduling prompter for individuals with dementia using video phone. Lecture Notes in Computer Science (including subseries Lecture Notes in Artificial Intelligence and Lecture Notes in Bioinformatics) 5614 LNCS (PART 1), 429–438 (2009). https://doi.org/10.1007/978-3-642-02707-949
62. Shannon, K., Bail, K., Neville, S.: Dementia-friendly community initiatives: an integrative review. J. Clin. Nurs. **28**(11–12), 2035–2045 (Jun 2019). https://doi.org/10.1111/jocn.14746
63. Kontos, P.C., Naglie, G.: Bridging theory and practice. Dementia 6(4), 549–569 (Nov 2007). https://doi.org/10.1177/1471301207084394

The manufacturer's authorised representative in the EU is Springer Nature Customer Service Centre GmbH, Europaplatz 3, 69115 Heidelberg, Germany. If you have any concerns regarding our products, please contact ProductSafety@springernature.com

Printed and bound by CPI Group (UK) Ltd, Croydon, CR0 4YY
28/09/2026
02224089-0020